Ionic Liquids - Thermophysical Properties and Applications

Edited by S. M. Sohel Murshed

Published in London, United Kingdom

IntechOpen

Supporting open minds since 2005

Ionic Liquids - Thermophysical Properties and Applications
http://dx.doi.org/10.5772/intechopen.91572
Edited by S. M. Sohel Murshed

Contributors
Umaima Gazal, Samir Ibrahim I. Abu-Eishah, Saber A.A. Elsuccary, Thikrayat H. Al-Attar, Asia A. Khanji, Hifsa P. Butt, Nourah M. Mohamed, Ana B. Pereiro, Nicole S. M. Vieira, Margarida L. Ferreira, Paulo J. Castro, João M. M. Araújo, Isiah M. Warner, Rocío L. Pérez, Caitlan E. Ayala, Patricia Iglesias Victoria, Hong Guo, Rina Wasserman

First published in London, United Kingdom, 2021 by IntechOpen
IntechOpen is the global imprint of INTECHOPEN LIMITED, registered in England and Wales, registration number: 11086078, 5 Princes Gate Court, London, SW7 2QJ, United Kingdom
Printed in Croatia

British Library Cataloguing-in-Publication Data
A catalogue record for this book is available from the British Library

Additional hard and PDF copies can be obtained from orders@intechopen.com

Ionic Liquids - Thermophysical Properties and Applications
Edited by S. M. Sohel Murshed
p. cm.
Print ISBN 978-1-83968-498-2
Online ISBN 978-1-83968-499-9
eBook (PDF) ISBN 978-1-83968-500-2

For EU product safety concerns:
IN TECH d.o.o., Prolaz Marije Krucifikse Kozulić 3, 51000 Rijeka, Croatia, info@intechopen.com or visit our website at intechopen.com.

Meet the editor

Prof. S. M. Sohel Murshed was born in Bangladesh and obtained his Ph.D. in Mechanical and Aerospace Engineering from Nanyang Technological University, Singapore. He is currently a professor in the Mechanical Engineering Department, University of Lisbon, Portugal, and a visiting professor at Rochester Institute of Technology, New York. Previously he worked as a postdoctoral fellow and visiting professor and scientist at different universities in Singapore, the United States, the United Kingdom, and India. In 2020, he received the prestigious DUO-India Professorial Fellowship Award. Dr. Murshed has authored/co-authored 10 books, 30 book chapters, and more than 180 papers in leading international journals and conferences. He was recently named one of the World's Top 2% Scientists by Stanford University.

Contents

Preface

Ionic liquids (ILs) have received increasing interest from researchers and industries due to their fascinating properties and great potential in numerous applications. The usages of ILs are expanding every day in areas such as engineering, analytics, physical chemistry, electrochemistry, tribology, and biology. These liquids are also considered sustainable and green liquids that can be tailored for specific applications. The thermophysical properties of ILs are essential for their sustainable and high performance in real-world applications. Over six chapters, this book examines the properties and applications of these emerging liquids.

Chapter 1 presents a new process combining experimental work and kinetic analyses to produce [Bmim][Ac] IL, which has low vapor pressure and is considered one of the important ILs in the solvent industry. In addition, the chapter reports on the production of silver chloride as a high-value chemical compound byproduct.

Chapter 2 reviews current research and progress in fluorinated ionic liquids (FILs) as task-specific materials. It also highlights the unique thermophysical and toxicological properties of compounds in addition to their application as task-specific materials in many fields of interest including biomedical applications and other engineering processes.

Chapter 3 reports an extensive review of the properties and application of a group of uniform materials based on organic salts (GUMBOS). Noting that ILs are special types (melting points below 100°C) of organic salts, the chapter focuses on recent developments and studies that provide fine-tuned and enhanced properties through transformation and recycling of diverse ionic compounds into solid-state ionic materials of greater utility.

New lubricants or lubricant additives with high performance and low toxicity are of great significance, particularly to reduce their negative impact on the environment. Chapter 4 reviews the current literature on the development and use of ILs such as protic ILs as high-performance lubricants and lubricant additives to different types of base lubricants. It also elaborates the relation between the structures of ILs and their various features and properties including viscosity, thermal stability, corrosion behavior, biodegradability, and toxicity. The chapter also discusses friction reduction and wear protection mechanisms of ILs.

Chapter 5 discusses the applications of ILs in gas chromatography. It emphasizes the use of different types of ILs in different stages (static) and phases (stationary) in gas chromatography. This chapter also highlights the potential of ILs in multidimensional gas chromatography.

The final chapter presents a historical and technological review of ancient and contemporary industries based on alkali and alkali–earth salts and hydroxides. This review of the archeological, historical, and technological background provides readers with the scope of the various daily life applications of these salts and hydroxides from ancient times to today. The review reveals that many modern chemical manufacturing processes using alkali and alkali–earth salts and hydroxides have an ancient history.

I would like to thank all the authors for their contributions and the staff at IntechOpen for their cooperation and support.

I dedicate this book to my schoolteachers in Maijpara who are the unsung heroes of my success (if any).

S.M. Sohel Murshed, Ph.D.
Instituto Superior Técnico,
Lisbon, Portugal

Rochester Institute of Technology,
New York, USA

Chapter 1

Production of 1-Butyl-3-Methylimidazolium Acetate [Bmim][Ac] Using 1-Butyl-3-Methylimidazolium Chloride [Bmim]Cl and Silver Acetate: A Kinetic Study

Samir I. Abu-Eishah, Saber A.A. Elsuccary, Thikrayat H. Al-Attar, Asia A. Khanji, Hifsa P. Butt and Nourah M. Mohamed

Abstract

Since most of the literature alternatives used to produce the ionic liquid 1-butyl-3-methylimidazolium acetate [Bmim][Ac] are very slow and require different solvents, we have used in this work a new process to produce the [Bmim][Ac] by the reaction of the ionic liquid 1-butyl-3-methylimidazolium chloride [Bmim]Cl with silver acetate (AgAc) where silver chloride (AgCl) precipitates as a by-product. The genuine experimental work and kinetic analyses presented here indicate that the reaction rate constant k = 7.67x10^{12} e$^{(-79.285/RT)}$. That is, the Arrhenius constant k_o = 7.67x10^{12} L/mol.s and the activation energy E_a = 79.285 kJ/mol. The very high value of the Arrhenius constant indicates that the reaction of [Bmim]Cl with silver acetate to produce [Bmim][Ac] and silver chloride is extremely fast.

Keywords: ionic liquids, production, 1-butyl-3-methylimidazolium acetate [Bmim][Ac], 1-butyl-3-methylimidazolium chloride [Bmim]Cl, silver acetate, silver chloride, kinetic study

1. Introduction

The last two decades has witnessed a growth in the research activities related to ionic liquids (ILs). Most of the work focus on replacing the widely used volatile organic solvents (VOCs) by suitable alternative solvents with minimum chemical waste and environmental pollution. The readily available VOCs have some ecological constraints such as high volatility, fire hazardous, risk explosion, and toxicity that force researchers to develop better and safer solvents.

In general, ILs are in liquid state at below 100 °C and possess negligible vapor pressure [1–4]. They have gained more applications nowadays as an important class of non-toxic, non-volatile, environmentally-friendly solvents in (bio)catalysis —applicable to many ionic, polar and nonpolar structure groups— and as efficient electrolytes [5]. In addition, ILs are good solvents for a wide range of inorganic and organic materials, have high thermal stability, high ionic conductivity and easy recyclability; these are some reasons to consider ionic liquids as "green solvents" [5]. The increased interest in ILs since 1990 is clearly due to the realization that these materials, formerly used for electrochemical applications including electrolytes for batteries, capacitors and charge storage devices as well as in the area of biomass utilization [6].

The ionic liquid of interest in this work is the 1-butyl-3-methylimidazolium acetate [Bmim][Ac], which has a low vapor pressure, hydrophilic, and is considered as one of the emerging important ILs in the solvent industry which has some promising applications as a solvent for lignin. This IL is produced as a reagent mainly in United States, Germany, France, and China. The current price of 1 kg of [Bmim][Ac] IL is about 785 EUROS [7, 8].

In this work, we will try to select the most feasible process alternative among the others to produce [Bmim][Ac] ionic liquid. Then to experimentally determine the kinetic data necessary to design a continuous stirred tank reactor (CSTR) for the production of [Bmim][Ac] based on the selected process, i.e. to determine the rate equation of the reaction and its order with respect to both reactants, the rate constant (k) as a function of temperature, and the activation energy of the reaction (E_a).

2. Uses of [Bmim][Ac]

There are several needs related to the [Bmim][Ac] ionic liquid and has many advantages over conventional organic solvents used nowadays due to it significantly low vapor pressure and relatively high solubility. Although [Bmim][Ac] is not a widely available product, it is preferred over other solvents in the extraction of lignin; the primary natural polymer found in wood [2].

Different ionic liquids, containing the $Bmim^+$ cation, are able to efficiently dissolve cellulose. However, the ability of ILs to truly dissolve cellulose is significant when cellulose derivatization is attempted. A series of experiments on etherification (carboxymethylation) of cellulose was performed by [9] using both the conventional suspension approach (slurry) with 2-propanol as the principal reaction media and a totally homogenous reaction approach using ionic liquids as a reaction media capable of dissolving cellulose.

Upon a totally homogenous etherification, the [Bmim][Ac] ionic liquid was found to give the highest degree of substitution. The product obtained was water-soluble and had a degree of substitution (DS) of 0.59. The substitution pattern of the products obtained from the homogenous reactions follow the same substitution pattern as the products obtained from the conventional suspension process. This indicates that the properties of the products are in line with products prepared via the conventional reaction route [9].

Low solubility and undesirable denaturation in conventional solvents still represent a significant challenge for efficient extraction, accurate characterization and multipurpose processing of collagen, which is important in fighting the visible effects of aging on the skin. [Bmim][Ac] was evaluated as an alternative solvent for type I collagen [10]. Real-time polarizing optical microscope observation indicated complete disintegration of hierarchical structure of collagen aggregates as solubilized in [Bmim][Ac] at 25 °C where the solubility reached 8.0 wt.%; > 10 times higher than that in conventional dilute acetic acid. The high solubility of collagen in [Bmim][Ac] at 25 °C is ascribed to the loose binding between $[Bmim]^+$ and

$[CH_3COO]^-$, as well as stronger proton-accepting ability of the [Bmim][Ac], which enabled rupture of those intermolecular hydrogen bonds and the ionic bonds that stabilized the collagen aggregates. However, such bond-rupturing effect was found selective at room temperature [10].

As demonstrated by various instrumental analyses, the [Bmim][Ac] did not destroy the special triple-helical structure of tropocollagen molecules that had been identified as being of importance for the functional and bioactive properties of collagen. According to these results, the discovery of [Bmim][Ac] as an ideal solvent for collagen may open up new possibilities for the chemistry and engineering of collagen, which has long been established as a readily accessible and renewable resource with many unique properties [10].

Preparation of amidoxime from nitriles in molecular solvent (usually in an alcohol) are accompanied by the amide side products. Surprisingly a selective formation of the desired amidoxime was observed in [Bmim][Ac] IL. No reaction occurred in imidazolium-based ionic liquids, containing other anions. The selectivity of the reaction was investigated for the preparation of a drug candidate's intermediate with similar result. Selective amidoxime formation in [Bmim][Ac] ionic liquid was proven for other model compounds too [11].

The internal redox esterification of α, β-unsaturated aldehydes and alcohols using different ionic liquids as catalysts and reaction solvents was carried out by [12] who found that the basic ionic liquid [Bmim][Ac] exhibited the best activity for this reaction.

Other applications of [Bmim][Ac] is in the biochemical industry where it can provide a strong addition to that industry as an ideal solvent for biomaterials involved in production processes, such as isolating lignin in paper pulp bleaching process, that provides an effective alternative to the conventional VOCs [13], Moreover, [Bmim][Ac] provides a useful extractor to separate collagen without destroying its intrinsic bioactive bonds when pure collagen is required as one of their ingredients.

3. Production of [Bmim][Ac]

There are several chemical paths to produce [Bmim][Ac], each of which can be considered as an alternative that requires certain design requirements mostly different from those required by the other alternatives. The anion exchange method can be used to produce water-soluble ionic liquids such as [Bmim][Ac] from reaction of halide ionic liquids such as [Bmim]Br, [Bmim]Cl, [Emim]Cl, etc. as a source of the anion and an acetate solution as a source of the acetate cation. The following is a summary of the several available paths for synthesis of [Bmim][Ac]:

1. An aqueous solution of 1-butyl-3-methylimidazolium bromide [Bmim]Br was allowed to pass through a column filled with anion exchange resin to obtain [Bmim][OH]. The aqueous [Bmim][OH] solution was then neutralized with equal molar acetic acid [CH_3COOH]. After removing water by evaporation under vacuum, the viscous liquid [Bmim][Ac] was thoroughly washed with diethyl ether, and finally dried under vacuum for 72 h at 70 °C [14].

2. Silver acetate (AgAc) (0.67 g, 4 mmol) was added to a solution of [Bmim]Cl (0.700 g, 4 mmol) in water (10 mL) and stirred at room temperature for 4 h. The suspension was filtered to remove silver chloride. The water was removed under vacuum to yield 0.69 g (85 wt.%) of a colorless oil [Bmim][Ac] [14].

3. Sodium 4-*tert*-butylphenolate (9.85 g, 57.25 mmol) was added to a solution of [Bmim]Cl (10 g, 57.25 mmol) in dry 2-butanone (500 mL). The reaction mixture was stirred vigorously for 12 h and afterwards filtered through Celite. An aqueous solution (500 mL) of acetic acid [CH_3COOH] (5.15 g, 85.87 mmol) was added then to the reaction mixture and stirred for half an hour. The organic phase was separated and washed with 50 mL of H_2O. The water was removed under vacuum to yield the product as a colorless liquid. The yield was 9.08 g (80 wt.%) [15].

4. [Bmim][Ac] can be synthesized by the addition of potassium acetate [CH_3COOK] to [Bmim]Cl IL in dry acetone. The formed KCl is insoluble and it precipitates and can be easily removed by filtration. But the main impurity in the [Bmim][Ac] product obtained by this method is the remaining [Bmim]Cl. In order to reduce this impurity, different amounts of [Bmim]Cl were added to [Bmim][Ac] and measured with an alkaline copper standard. Recovery rates of 90–98% [Bmim][Ac] was obtained over the whole range of the [Bmim]Cl concentration [16].

5. [Bmim][Ac] can be synthesized by dissolving [Bmim]Cl (0.25 mmol) in dry acetone (50 ml) and stirred with (5.0 g) ammonium acetate (CH_3COONH_4) at room temperature for 24 h to exchange the anion. The reaction mixture was then filtered off to remove precipitated ammonium chloride [NH_4Cl] and the excess ammonium acetate. The acetone was evaporated on rotary evaporator under reduced pressure and dried under vacuum to yield 96 wt.% [Bmim][Ac] [17].

6. [Bmim][Ac] can be synthesized by the addition of silver acetate [CH_3COOAg] (0.67 g, 4 mmol) to a solution of [Bmim]Cl (0.700 g, 4 mmol) in water (10 mL) and stirred at room temperature for 4 h. The suspension was filtered to remove the silver chloride precipitate by-product. The water was removed under vacuum to yield 0.69 g (85 wt.%) of a colorless [Bmim][Ac] IL [18].

The [Bmim][Ac] can also be synthesized by the slow addition of acetic acid [CH_3COOH] (10 mL, 180 mmol) to a 30 wt.% methanol solution of 1-Butyl-3-methylimidazolium methyl carbonate [Bmim][$MeCO_3$] (140 mL, 175 mmol) and stirred for 1 h under a dynamic vacuum (Schlenk line) to obtain [Bmim][Ac] (33.072 g; 167 mmol; 95% yield), which was further dried on a Schlenk line for 48 hours at 60 °C [16].

4. Alternatives processes for production of [Bmim][Ac]

In this work, we have qualitatively prioritized three different alternatives to provide a basis that helps in selecting the most suitable process alternative among the others to produce [Bmim][Ac] ionic liquid. These three process alternatives are discussed below.

Alternative 1: Butylation of 1-imidazole and methylation of 1-butylimidazole using Packed Bed Reactors (PBRs)

Here we have three main reactions as shown in the reaction schemes below: (1) Butylation of Imidazole by butyl iodide to produce Butylimidazole, (2) Methylation of the Butylimidazole by di-methyl carbonate to produce 1-Buty-3-methylimidazolium ion and acetate counter ion, (3) Ion-exchange reaction of the resulting

1-Buty-3-methylimidazolium ion and acetate counter ion to 1-butyl-3-methylimidazolium acetate [Bmim][Ac] in presence of excess acetic acid. See **Figure 1**.

Heating is required to bring the temperature of the first and second reactions to 150 °C and 210 °C, respectively. The third reaction is run at 80 °C. However, the second reaction must be operated at very high pressure (>70 bar), which is a special concern that requires very thick-wall equipment and further safety considerations. The reaction residence times for the first and the second reactions are 5 hr. and 2 hr., respectively. This alternative also uses Al_2O_3 catalyst to increase the reaction rate and decrease the residence time. However, using a catalyst increases the process cost; thus, it must be justified, especially if the reaction time is still high.

Alternative 2: Methylation of 1-butylimidazole using Packed Bed Reactors (PBRs)

Here we have two main reactions as shown in the reaction schemes below: (1) Methylation of Butylimidazole by dimethyl carbonate to produce 1-Buty-3-methylimidazolium ion and acetate counter ion, (2) Ion-exchange reaction of the resulting 1-Buty-3-methylimidazolium ion and acetate counter ion to 1-butyl-3-methylimidazolium acetate [Bmim][Ac] in presence of excess acetic acid. See **Figure 2**.

Heating is required to bring the temperature of the first and second reactions to 210 °C and 80 °C, respectively. However, the heating requirements here is less than that in Alternative 1. As in Alternative 1, high pressure (>70 bars), vacuum distillation, and use of Al_2O_3 catalyst, need to be considered in this alternative too. However, this alternative requires less time for the first reaction, which is, reduced from 5 to only 2 hours.

Alternative 3: Butylation of 1-methylimidazole using Micro-Structured Reactor (MSR)

Here we have two main reactions as shown in the reaction schemes below: (1) Butylation of Methylimidazole by 1-Chlorobutane to produce [Bmim]Cl. See for

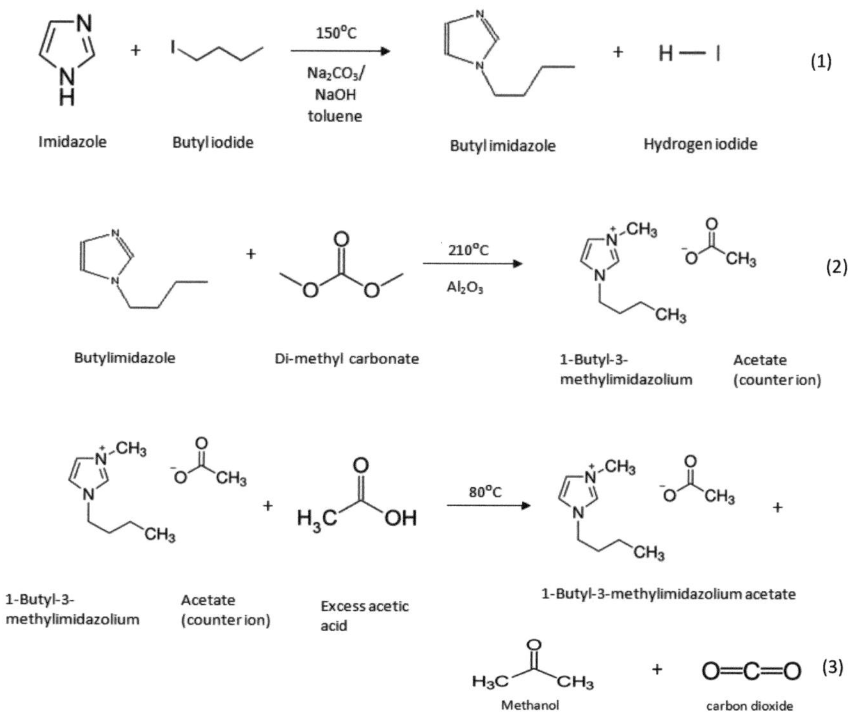

Figure 1.
Butylation of 1-imidazole and methylation of 1-butylimidazole using packed bed reactors (PBRs).

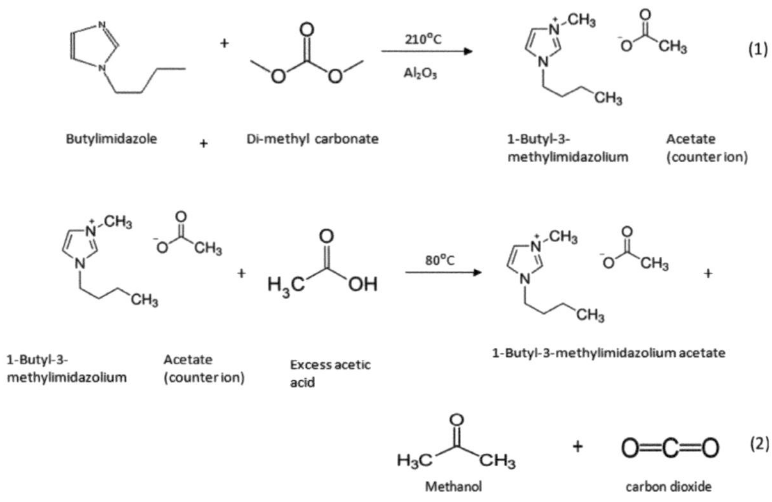

Figure 2.
Methylation of 1-butylimidazole using packed bed reactors (PBRs).

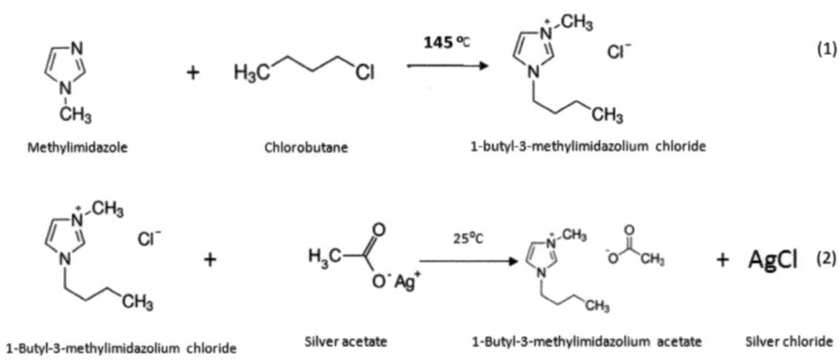

Figure 3.
Butylation of 1-methylimidazole using micro-structured reactor (MSR).

example, [14, 19, 20], (2) Ion-exchange reaction of the resulting [Bmim]Cl with silver acetate to produce 1-butyl-3-methylimidazolium acetate [Bmim][Ac] [21]. See **Figure 3**.

In this alternative, heating is only required in the first reaction to 145 °C where the pressure is around 6 bar. Also, the second reaction is operated at or near atmospheric pressure. This is a major advantage for this alternative where safety considerations and cost are dramatically reduced. The residence time for the first reaction is also relatively short (~32 min) at which about 87% conversion of the reactants is achieved when the reaction is carried out in a Micro-Structured Reactor (MSR) [22], which is definitely a great advantage for this alternative. The residence time for the second reaction is only few seconds if carried near room temperature. Another advantage of this alternative is that it does not require any catalyst in either reaction.

5. Comparison of [Bmim][Ac] production alternatives and process selection

In order to select the best process for commercial production of [Bmim][Ac] among the above three developed alternatives, a logical comparison procedure has been followed based on the following main criteria: Safety and environmental criterion, preliminary economic feasibility criterion, operating conditions criterion, and process complexity criterion. Hence, a number of comparison tables were developed to give a clear picture about each of these alternatives and enable us in selecting the most promising alternative among the others.

Safety and environmental concerns criterion

Safety and protection of the environment are intrinsic considerations that should be focused on when designing a plant since for any success of the manufacturer, it is important that the personnel working in the industry and the environment surrounding it remains safe and complies with the nation's environmental regulations. In Alternative 1, high number of chemicals are involved in the process (see **Table 1**); most of which are flammable and combustible, i.e. might form explosive vapor mixtures and ignite near the source. Alternative 2 has almost the same number and type of chemicals (except imidazole) as in Alternative 1. Alternative 3 has only 4 chemicals; only two of which are flammable. Thus, Alternative 3 is considered to be the most environmentally-friendly and safe process among the studied three alternatives.

Preliminary economic feasibility criterion

Economic feasibility study is considered the first step in calculating and estimating the expected cost and profit for an industrial process. Hence, it enables the early evaluating for the cost and estimated profit for different alternatives. The preliminary economic feasibility is one important criterion used to evaluate the process production alternatives. It is a preliminary indication of the project's profitability, which is calculated by subtracting the cost of raw materials from the price of the final product [Bmim][Ac], according to the following definition:

$$\text{Preliminary economic feasibility} = \text{Price of [Bmim]Ac} - \sum \text{Costs of Reactants}$$

Chemical	Alternative	Safety and environmental concerns
Imidazole	1	May be combustible at high temperature [23]
Butyl iodide	1	Flammable liquid (Class 3). Vapors may form an explosive mixture with air [24]
Dimethyl carbonate	1 & 2	Highly flammable liquid; Flash point = 18 °C [25]
Acetic acid	1 & 2	Flammable in presence of open flames and sparks of heat. Ecotoxicity in water = 423 mg/L [26]
Hydrogen iodide	1	Non-flammable gas but hydrolyzes very rapidly yielding hydroiodic acid [27]
Methanol	1 & 2	Volatile and flammable. It may be slightly toxic to aquatic life
Toluene	1	Flammable
Butyl imidazole	2	Combustible: may burn but does not ignite readily. Flash point = 110 °C [28]
Methyl imidazole	3	Combustible [29]
1-Chlorobutane	3	Flammable liquid (Class 3). Low toxicity to aquatic organisms [30]
Silver acetate	3	Non-flammable. Toxic to aquatic organisms, may cause long-term adverse effects in the aquatic environment [31]
Silver chloride	3	Non-flammable. Does not pose adverse effect on aquatic life [32]

Table 1.
Safety and environmental concerns of the chemicals involved in the three alternatives.

Table 2 shows the individual chemicals prices in 2020 while **Table 3** shows the cost of reactants, the expected price for the sellable products and the difference between cost and sellable price for the desired product. **Table 2** also shows that Alternative 3 has the highest positive difference according to the above definition, and hence has the highest expected profit.

Process operating conditions criterion

Process operating conditions (pressure, temperature, reaction time, etc.) usually affect process selection, design and its economy since dealing with unfavorable conditions may raise safety concerns and increase process capital and operating costs (and thus process profitability). **Table 4** summarizes the process conditions for each of the studied alternatives. It is clear from **Table 4** that Alternative 3 can be

Compound	2020 Price (Euro, €)
1-Butyl-3-methylimidazolium acetate	980/kg
Imidazole	169/500 g
1-Butylimidazole	87.1/100 g
1-methylimidazole	141/500 g
Butyl iodide (1-Iodobutane)	125/500 g
Dimethyl carbonate	257/2 L
1-Chlorobutane	138/L
Aluminum oxide	137/kg
Acetic Acid	120/2.5 L
Methanol	73.7/L
Silver acetate	646/100 g
Silver chloride	5110/kg

Table 2.
Individual chemicals prices in 2020 [8].

Item	Alternative 1*	Alternative 2	Alternative 3
Reactants (and reagents)	Imidazole, Butyl iodide, Dimethyl carbonate, Al_2O_3 (catalyst), Acetic acid	Butyl imidazole, Dimethyl carbonate, Al_2O_3 (catalyst), Acetic acid	Methyl imidazole, 1-Chlorobutane, Silver acetate
Sellable Product(s)	[Bmim][Ac]	[Bmim][Ac]	[Bmim][Ac], AgCl
Cost of Reactants (Euro, €)	757.45	1061.07	2535.81
Price of Sellable Product (s) (Euro, €)	2765.00	2765.00	4933.38
Preliminary economic feasibility (Euro, €)	2007.55	1703.93	2397.57
Preliminary economic feasibility (US $)	2543.50	2158.80	3037.60

**The cost of NaOH, Na_2CO_3 and toluene are not counted in the preliminary economic feasibility.*

Table 3.
Preliminary economic feasibility results for the three alternatives studied in this work based on the raw materials' and final product(s)' prices.

Item	Alternative 1	Alternative 2	Alternative 3
Reactants	Imidazole, Butyl iodide, Dimethyl carbonate, Acetic acid	Butyl imidazole, Dimethyl carbonate, Acetic acid	Methyl imidazole, 1-Chlorobutane, Silver acetate
By-product(s)	CO_2, methanol, HI	CO_2, methanol	Silver chloride
Catalysts involved	Al_2O_3	Al_2O_3	None
Others	NaOH, Na_2CO_3, Toluene	None	None
Main reactions' temperatures (°C)	150, 210 and 80	210, 80 and 80	154 and room temperature
Main reaction pressure (bar)	70	70	6
Main reaction(s)' residence time	5 hr., 2 hr	7 hr	31.7 min, few seconds

Table 4.
Comparison of [Bmim][Ac] production alternatives in terms of reactants involved, products obtained, operating temperature, operating pressure, etc.

conducted at 145 °C and 6 bar for the first reaction and at near room temperature and 1 bar for the second reaction, which are much lower those required for Alternatives 1 and 2. Also Alternative 3 has the most favorable residence time (31.7 min for the first reaction and few seconds for the second reaction) when compared with those required for Alternatives 1 and 2. In addition, no catalyst is required for Alternative 3, while Al_2O_3 catalyst is required in the other two alternatives. So, one can say, Alternative 3 has the most favorable operating conditions among the three Alternatives studied.

Process complexity criterion

Since there are many compounds involved in Alternatives 1 and 2, the complexity of a process increases since more reaction and separation steps are needed and hence the capital and operating cost will dramatically increase. **Table 4** above shows Alternative 3 has the least number of compounds involved, thus it is the least complex alternative.

Thus, based on the analyses presented in **Tables 1–4** above, Alternative 3 has the highest preliminary feasibility and the most favorable operating conditions, the least process complexity and the minimum environmental and safety concerns.

6. Experimental setup, procedure and software used

Since most of the above methods are slow and require different solvents, the silver acetate [AgAc] method is used in this work to produce [Bmim][Ac] according to the following reaction:

$$[\text{Bmim}][\text{Cl}] + \text{AgAc} \rightarrow [\text{Bmim}][\text{Ac}] + \text{AgCl} \quad (1)$$

Or,

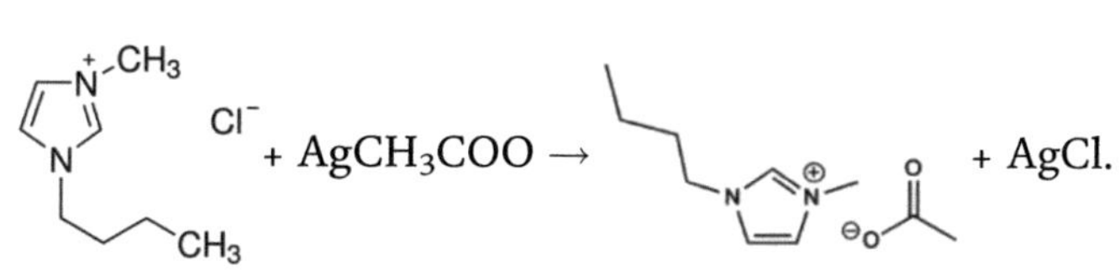

Or,

$$A + B \rightarrow C + D$$

Here, a silver chloride by-product is produced that can compensate for the cost of the silver acetate raw material. The experiment to produce the ionic liquid [Bmim][Ac] according to Eq. (1) was carried out in a CSTR.

The information available from literature [18] about this reaction are as follows: the conversion and reaction time at 25 °C are 84% and 4 h, respectively, when the ratio between [Bmim]Cl and silver acetate is 1:1.

As per the fact that ionic liquids are relatively newly researched species, their chemical analysis is of limited methods. Hence, from the reaction equation, one can notice that the only product that could be analyzed to follow up the progress of the reaction is AgCl. The Ag ions have some very common methods of determination such as titration or the most extensively used method gravimetric analysis. However, for the purpose of this experiment, sequential trials using the above-mentioned methods is time consuming and impractical considering the limited amount of precipitate. More about gravimetric analysis can be found elsewhere [33].

Noteworthy, the kinetics of the reaction could only be measured through the following up of the decrease in the concentration of Ag and/or Cl ions that could be easily monitored by the potentiometric detection technique based on ion-selective electrode. The potentiometric detection technique, as a simple method, offers several advantages such as speed and ease of preparation and procedures, simple instrumentation, relatively fast response, wide dynamic range, reasonable selectivity, application in colored and turbid solutions and low cost.

In this experimental work, a silver sheet coated with AgCl served as a working electrode and the reference electrode was a Jenway Ag/AgCl double junction containing 1.0 mol/L of lithium acetate solution in the outer compartment (shown in **Figure 4**). The cell potential was measured using a one-channel high-input impedance module (HIM) [34] attached to ADC-20 data acquisition card (purchased from Pico Technology Limited, London, UK) connected to a personal computer (PC). The potential was continuously output to the PC through the PicoLog recorder software. The electrochemical cell may be represented as follows: Ag/AgCl(s)/sample solution/1.0 mol/L CH_3COOLi salt bridge/4.0 mol/L KCl/Ag/AgCl.

In this work, a newer, more sophisticated method of monitoring Ag ion concentration was chosen which is known as data acquisition method. The system control is maintained through a data logging software which uses electrodes to detect the potential difference of the solution with time. This method was chosen here because

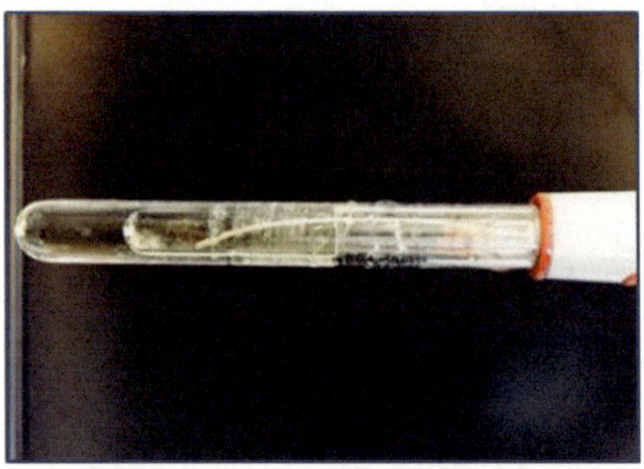

Figure 4.
Jenway Ag/AgCl double junction reference electrode.

it is fast and produce accurate results, it requires minimal monitoring and it is reliable to be used for a large number of runs. The few limitations associated with this method is the need for calibration for each run to get the unique relationship between the potential difference and concentration.

The main instrument used for the data acquisition was the Picolog® high-resolution data logger from Pico Technology [35]. It allows the experimenter to achieve fast and reliable results due to its ability to detect small changes. Also, ease of manipulating and displaying of data makes this particular setup a useful component to have numerous readings at a predetermined sampling rate. It is also powered directly by the PC connection and does not require external batteries or power source [35].

7. Generation of Ag^+ concentration calibration curves

Since the analytical technique to measure the Ag ion concentration does not measure the concentration directly, a calibration curve and subsequently a calibration equation is required to form the relationship between the signal, which is the potential in millivolts (mV), and the molar concentration (M). The complete setup is shown in **Figure 5**.

The experimental procedure used in this work is as follows:

1. To a 100 mL double walled beaker, add 50 mL of water and 5 mL of potassium acetate (x M) for the purpose of adjusting the ionic strength and obtaining a steady baseline.

2. Fill the circulating water bath with ice water to keep a low temperature since a test run of the experiment at room temperature indicated that the reaction was very fast and therefore to better study the reaction kinetics, a temperature < 25 °C was used.

3. Prepare three standard solutions of silver acetate (AgAc) with molar concentrations of 0.0001 M, 0.001 M and 0.01 M.

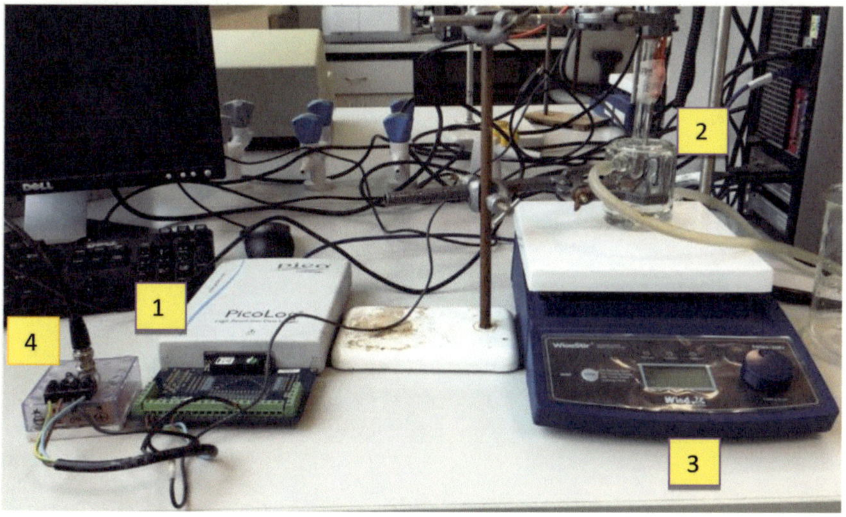

Figure 5.
Experimental setup; 1: PicoLog data logging device, 2: Glass beaker, 3: Stirring plate, 4: High-input impedance module (HIM).

4. Sequentially add appropriate small aliquots of AgAc standards and record the potential (mV) continuously. A final Ag ion concentration range of ($2.37x10^{-6}$ – $5.14x10^{-4}$ M) was tested to check the Nernstian response of the working electrode in order to select a reasonable initial AgAc concentration in the subsequent reaction kinetics tests.

5. Generated the graph of the potential signal (mV) vs. the AgAc molar concentration (M).

For Run #1 (at T = 12 °C), the calibration curve data are shown in **Table 5** and **Figure 6**, and the plot of the calibration curve is shown in **Figure 7**. The experiment was run at the same temperature at which the calibration was performed, and therefore for any subsequent runs at different temperatures, a different calibration curve is required.

AgAc concentration (M)	2.37E-06	4.72E-06	2.82E-05	5.14E-05	2.84E-04	5.14E-04
Potential (mV)	240.6	248.6	255.3	258.4	299.2	337.1

Table 5.
Calibration curve data at T = 12 °C.

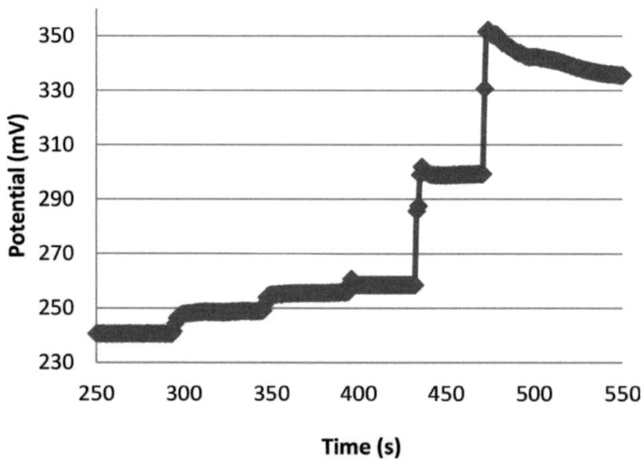

Figure 6.
The trend of the potential difference versus time for run #1 at T = 12 °C).

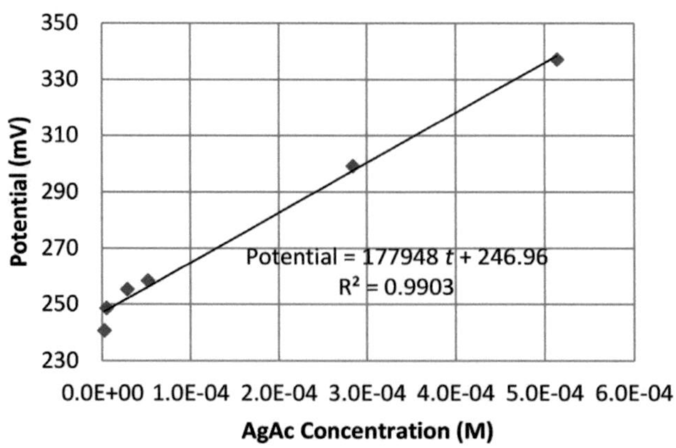

Figure 7.
Calibration-curve linear fit of the potential (mV) vs. AgAc molar concentration (M) for run #1 (at 12 °C).

8. Experimental results and analyses

The main objective of this experiment is to determine the kinetic data necessary to design a continuous stirred tank reactor (CSTR) for the production of 1-butyl-3-methylimidazolium acetate [Bmim][Ac] from the reaction of 1-butyl-3-methylimidazolium chloride [Bmim]Cl and silver acetate (AgAc), i.e. to determine the rate equation of the reaction and its order with respect to both reactants, the reaction rate constant (k) as a function of temperature, and the reaction activation energy (E_a). Several experimental runs for the reaction presented by Eq. (1) have been carried out. The purpose of each of these tests is also outlined below.

Run #1: Excess reactant method (isolating [Bmim]Cl) for the determination of the partial orders of the reactants.

A pseudo-first order reaction is a reaction where one of the reactants is present in large excess compared to the other reactant such that its concentration does not change significantly with time. In this case, the concentration of the excess reactant, say A, can be assumed to be constant and is absorbed into the rate constant k to give a pseudo-first order rate constant $k' = k\, C_A$. So, for the reaction presented by Eq. (1), $C_A >> C_B$, then $\Delta C_A \approx 0$ [36].

In the same way, some second and higher-order reactions can be more easily examined when the concentration of one reactant is essentially held constant (by using a large excess of that reactant) such that the fractional change in its concentration over the course of reaction is negligible [37].

Here Run #1 was carried out at 12 °C using excess of reactant A (i.e. [Bmim]Cl). The rate equation for the reaction presented by Eq. (1), is given by

$$-r_A = -\frac{dC_A}{dt} = (k\, {C_A}^{\alpha})\, {C_B}^{\beta} = k' {C_B}^{\beta} \tag{2}$$

where k is the reaction rate constant and k' is the reaction rate constant in presence of excess A (i.e. [Bmim]Cl). Here B stands for the silver acetate [AgAc]. Since

$$-r_A = -\frac{dC_A}{dt} = -\frac{dC_B}{dt} = k' {C_B}^{\beta} \tag{3}$$

By integration of Eq. (3), we get

$$\ln\left(-\frac{dC_B}{dt}\right) = \ln k' + \beta \ln C_B \tag{4}$$

The potential vs. time and reactant B (i.e. AgAc) concentration vs. time are shown in **Figures 8** and **9**, respectively. Both curves are straight lines with $R^2 \approx 1.0$.

From **Figure 9**, the AgAc concentration, $C_B = -0.000459\, t + 0.0001867$, thus $-\frac{dC_B}{dt} = -0.000459$, or $\frac{dC_B}{dt} = 0.000459$, i.e. it is constant. Thus, the plot of $\ln\left(-\frac{dC_B}{dt}\right)$ versus $\ln C_B$ will be a horizontal line with a zero slope. Accordingly, $\beta = 0$ and the reaction rate is of zero order with respect to the AgAc concentration.

Run # 2: Using the equimolar method for the determination of the partial orders of the reactants.

In order to determine the overall order of a chemical reaction, it is more convenient to use equimolar concentrations of the reactants A and B at the start of the reaction (i.e. $t = 0$) [38]. So, for the reaction presented by Eq. (1), and at any time t, the [Bmim]Cl concentration is equal to the AgAc concentration, or $C_A = C_B = C_0 - x = C$, where x is the reacted mole fraction of either component.

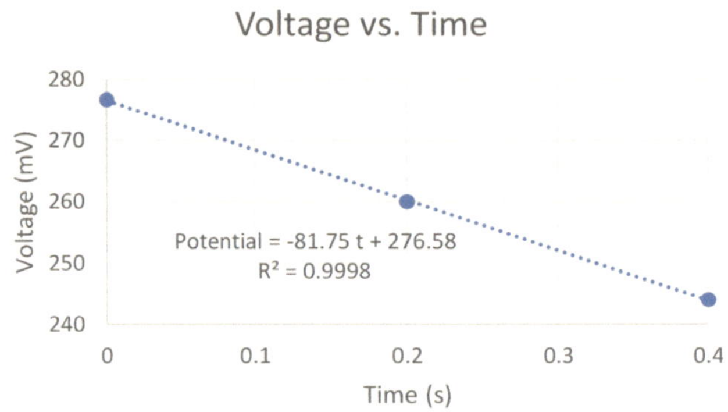

Figure 8.
Potential (mV) vs. time curve for run #1 at T = 12 °C.

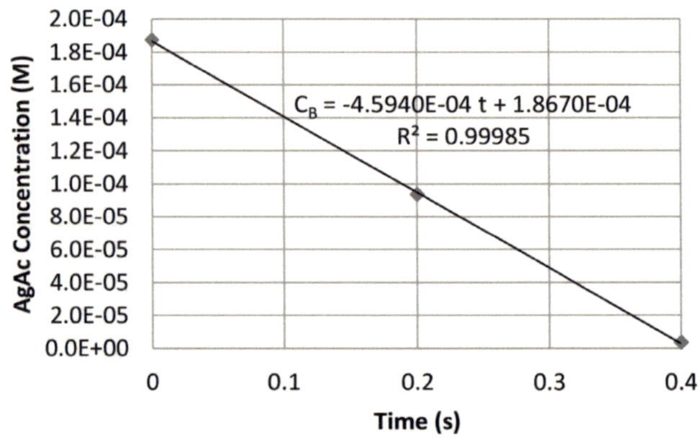

Figure 9.
AgAc concentration vs. time for run #1 at T = 12 °C.

In this case, Run # 2 was carried out at 12 °C using equimolar amounts of [Bmim]Cl and AgAc. The rate equation in this case can be rewritten as:

$$-r_A = -\frac{dC_A}{dt} = -\frac{dC_B}{dt} = k\ {C_A}^{\alpha+\beta} \tag{5}$$

where β was found to be zero (earlier in Run #1 results) when [Bmim]Cl was used in excess. Here we have two options for the C_A exponent (either α = 1 or $\alpha \neq 1$).

For α = 1, Eq. (5) can be rearranged to give

$$-\frac{dC_A}{C_A} = k\ dt \tag{6}$$

By integration of Eq. (6), we get:

$$-\int_{C_{Ao}}^{C_A} \frac{dC_A}{C_A} = \int_0^t kdt \tag{7}$$

Or,

$$\ln\left(\frac{C_A}{C_{Ao}}\right) = -k\ t \tag{8}$$

Or,

$$C_A = C_{Ao}\, e^{-kt} \tag{9}$$

The AgAc molar concentration (M) and the corresponding ln (C_{Ao}/C_A) vs. time for Runs #2 are given in **Table 6**. The plot of ln (C_{Ao}/C_A) vs. time is shown in **Figure 10**; from which $C_A = C_{Ao}\, e^{-kt} = C_{Ao}\, e^{-0.02079\, t}$. Here, $R^2 = 0.9425$, which means that the ln (C_{Ao}/C_A) vs. time is almost linear ($\alpha \approx 1.0$) and a first-order [Bmim]Cl concentration is a valid assumption.

For $\alpha \neq 1$, Eq. (7) becomes

$$-\int_{C_{Ao}}^{C_A} \frac{dC_A}{C_A^{\alpha}} = \int_0^t k dt \tag{10}$$

and the solution of Eq. (10) can be written as

$$\frac{C^{1-\alpha} - C_0^{1-\alpha}}{\alpha - 1} = k\, t \tag{11}$$

However, several attempts have been made in this work to find the non-integer value of α based on Eq. (11). In all runs and at all tested temperatures, the value of α was ≈ 1.0, which means that the first-order [Bmim]Cl concentration is still a valid assumption.

Now, in order to determine the *k* value as a function of temperature, two more runs have been conducted at 37.6 **°C** and 50 **°C**. The results are displayed below.

Time (s)	AgAc concentration (M)	Potential (mV)	ln (C_{Ao}/C_A)
0	0.0000818	257.919	0.0
1	0.0000794	257.486	0.02978
2	0.0000778	257.199	0.05014
3	0.0000769	257.042	0.06177
4	0.0000758	256.848	0.07618

Table 6.
AgAc molar concentration (M) and ln (C_{Ao}/C_A) vs. time for runs #2 at 12 °C.

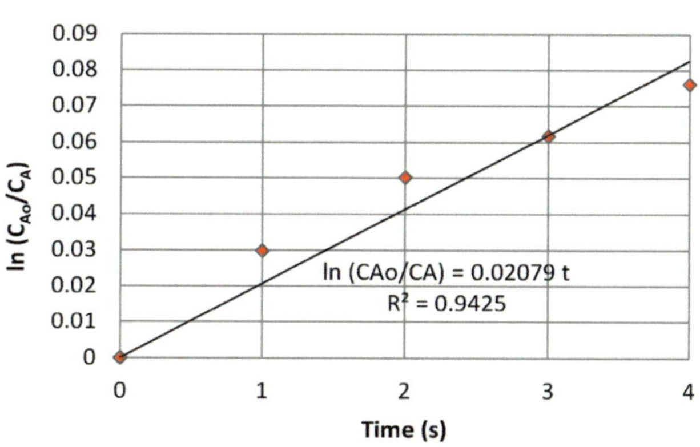

Figure 10.
Ln (C_{Ao}/C_A) vs. time for run #2 at 12 °C. ◆: Exp, ___: linear fit.

Run #3: Reaction kinetics at T = 37.6 °C.

This run was carried out using equimolar concentrations of the reactants A and B at the start of the reaction. The calibration curve data for this run are given in **Table 7**. The corresponding plots of the potential (mV) vs. AgAc molar concentration (M) and ln (C_{Ao}/C_A) vs. time are shown in **Figures 11** and **12**, respectively. Here, R^2 = 0.9443, and the ln (C_{Ao}/C_A) vs. time is almost linear ($\alpha \approx 1.0$) and is first order with respect to the [Bmim]Cl concentration. As seen from **Figure 12**, the rate constant at 37.6 °C is 0.50865 s^{-1}.

Run #4: Reaction kinetics at 50 °C.

Again, this run was carried out using equimolar concentrations of the reactants A and B at the start of the reaction. The calibration curve data for this run are given in **Table 8**. The corresponding plots of the potential (mV) vs. AgAc concentration (M) and ln (C_{Ao}/C_A) vs. time are shown in **Figures 13** and **14**, respectively. Again, R^2 = 0.9671 ≈ 1.0, and the ln (C_{Ao}/C_A) vs. time is almost linear ($\alpha \approx 1.0$) and first order with respect to the [Bmim]Cl concentration. As seen from **Figure 14**, the rate constant at 50 °C is 0.92047 s^{-1}.

From the linear fits of ln (C_{Ao}/C_A) vs. time at the test temperatures 12, 37.6 and 50 °C, **Table 9** shows the rate constant (k) values vs. temperature.

Lastly, **Figure 15** shows the linear fit plot of ln k vs. $1/T$. That is

$$\ln\,(k) = \ln\,(k_o) - \left(\frac{E_a}{R}\right)\frac{1}{T} \tag{12}$$

Time (s)	AgAc concentration (M)	Potential (mV)	ln (C_{A_o}/C_A)
0	0.000283	321.8	0.0
1	0.000239	314.6	0.17030
2	0.000105	292.9	0.99030
3	0.000417	282.6	1.91423
4	0.000374	281.9	2.02320
5	0.000257	280.0	2.39790

Table 7.
Calibration curve data for run #3 at T = 37.6 °C.

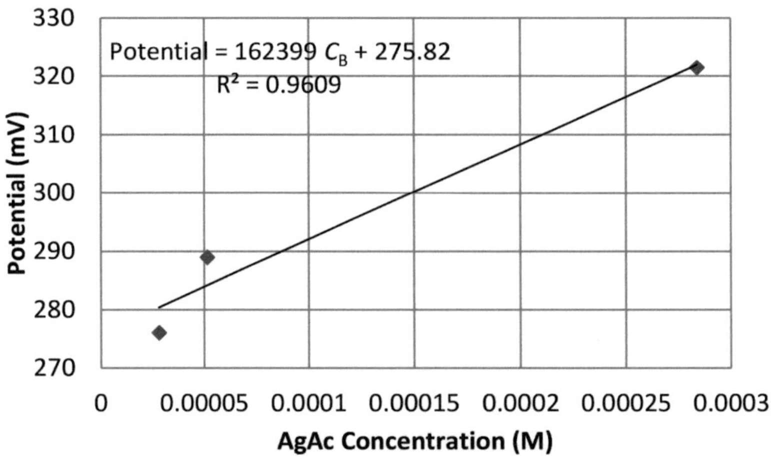

Figure 11.
Calibration-curve for potential (mV) vs. AgAc molar concentration (M) for run #3 at 37.6 °C.

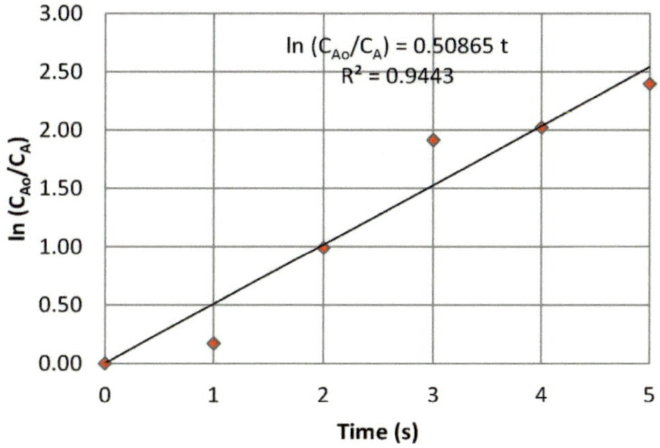

Figure 12.
Ln (C_{Ao}/C_A) vs. time for run #3 (at T = 37.6 °C); ◆: Exp, ___: linear fit.

Time (s)	AgAc concentration (M)	Potential (mV)	ln (C_{Ao}/C_A)
0	2.27E-04	299	0
0.2	1.97E-04	295.5	0.14085
0.4	1.72E-04	292.5	0.27971
0.6	1.42E-04	289	0.4706
0.8	1.08E-04	285	0.74555
1	8.31E-05	282.1	1.00617

Table 8.
Calibration curve data for run #4 at T = 50 °C.

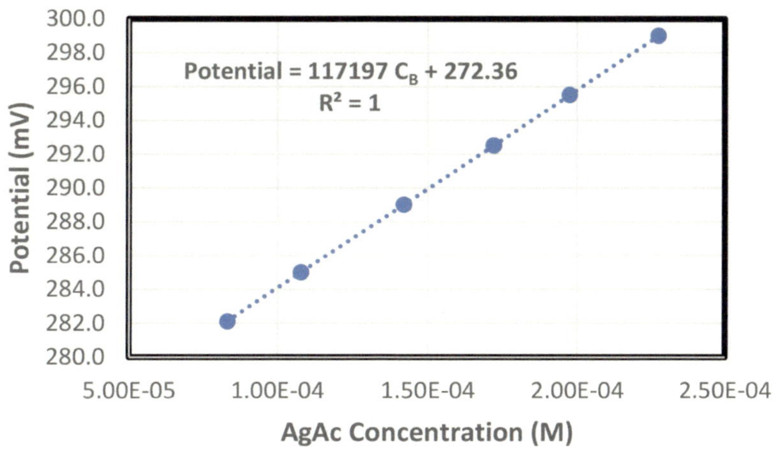

Figure 13.
Calibration-curve for potential vs. AgAc molar concentration for run #4 at 50 °C.

Figure 15 indicates that the relationship between ln k and $1/T$ is almost linear with R^2 = 0.9776. However, using the fitting parameters shown on **Figure 15**, the Arrhenius constant k_o and the activation energy E_a are determined as follows:

$$k_o = e^{29.668} = 7.67 \times 10^{12} \mathrm{L/mol.s.}$$

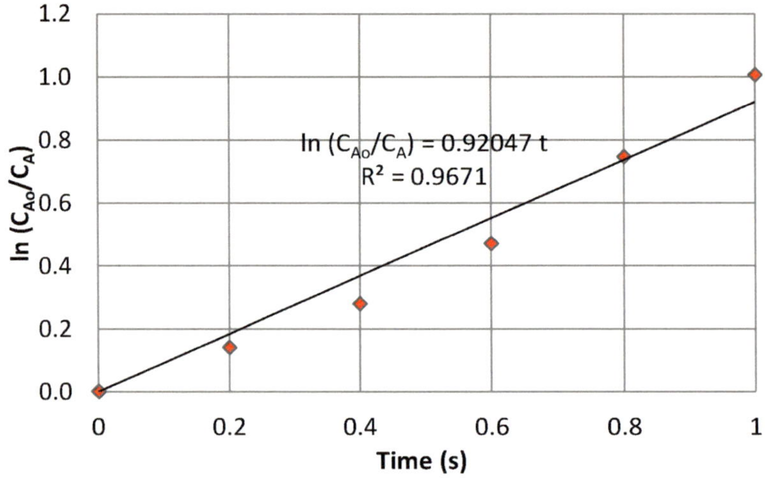

Figure 14.
Ln (C_{Ao}/C_A) *vs. time for run #4 (at* T = 50 °C*); ◆: Exp, ___: Linear fit.*

T (°C)	12	37.6	50
T (K)	285.15	310.75	323.15
k (s^{-1})	0.02079	0.50865	0.92047

Table 9.
Rate constant k *vs.* T *for the [Bmim][Ac] production reaction presented by Eq. (1).*

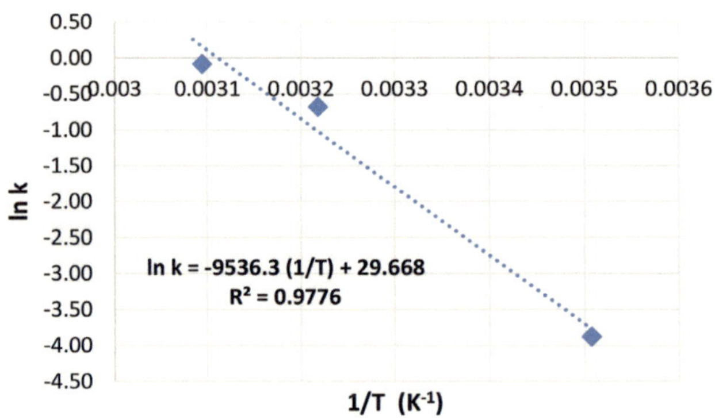

Figure 15.
Plot of ln k *vs.* 1/T *for [Bmim][Ac] production for the reaction presented in Eq. (1).*

$$E_a = 8.314 \times 9536.3/1000 = 79.285\text{kJ/mol}.$$

Finally, Eq. (12) can be written as.

$$k = k_o e^{(-Ea/RT)} = 7.67 \times 10^{12} e^{(-79.285/RT)} \tag{13}$$

Here, the Arrhenius constant k_o is extremely high, which means that the reaction of [Bmim]Cl and silver acetate to produce [Bmim][Ac] and silver chloride is extremely fast.

9. Conclusion

In this work, the kinetic data for the reaction of [Bmim]Cl and silver acetate to produce [Bmim][Ac] and silver chloride, were experimentally determined. The order of the reaction was found to be of first order with respect to [Bmim]Cl and of zero order with respect to silver acetate. The rate constant as a function of temperature was found to be k = 7.67x10^{12} e$^{(-79.285/RT)}$. That is, the values of k_o and E_a are 7.67x10^{12} L/mol.s and 79.285 kJ/mol, respectively. This indicates that the [Bmim]Cl reaction with silver acetate to produce [Bmim][Ac] and silver chloride is extremely fast. It should be mentioned here that the produced silver chloride has a very high-market value that can easily compensate for the high-initial cost of the silver acetate reactant.

Author details

Samir I. Abu-Eishah[1]*, Saber A.A. Elsuccary[2], Thikrayat H. Al-Attar[1], Asia A. Khanji[1], Hifsa P. Butt[1] and Nourah M. Mohamed[1]

1 Chemical and Petroleum Engineering Department, United Arab Emirates University, Al Ain, UAE

2 Chemistry Department, United Arab Emirates University, Al Ain, UAE

*Address all correspondence to: s.abueishah@uaeu.ac.ae

References

[1] Cull SG, Holbrey JD, Vargas-Mora V, Seddon KR, Lye GJ. Room-temperature ionic liquids as replacements for organic solvents in multiphase bioprocess operations, Biotechnol Bioeng. 2000; 69 (2):227–33.

[2] Bogolitsyn KG, Skrebets TE, Makhova TA. Physicochemical properties of 1-butyl-3-methylimidazolium acetate, Russian J General Chemistry. 2009;79:125–128.

[3] Sing G, Kumar A. Ionic Liquids: Physico-chemical, solvent properties and their applications in chemical processes, Indian J Chem. 2008;47A: 495–503.

[4] Azman AM (15 November 2006) Ionic Liquids in Organic Synthesis, Available at: https://docplayer.net/37205151-Ionic-liquids-in-organic-synthesis.html

[5] Wasserscheid P, Welton T. Ionic Liquids in Synthesis, Wiley-VCH Verlag GmbH & Co. KGaA, Weinheim, Germany, 2003. pp. 26–27.

[6] Sigma-Aldrich, editor (2005) ChemFiles: Enabling Technology, Ionic Liquids, Vol. 5.

[7] Solvionic (2003), Cleaner Solvent for Sustainable Chemistry, Available at: http://en.solvionic.com/products/1-butyl-3-methylimidazolium-acetate-98

[8] Sigma-Aldrich (2020), "[Bmim][Ac]", Available at: https://www.sigmaaldrich.com/catalog/search?term

[9] Mikkola J-P, Tuuf J-C, Kirilin A, Damlin P, Salmi T. Ionic liquid-aided carboxymethylation of Kraft pulp, Int. J. Chemical Reactor Eng. 2010;8(1):1542-6580, DOI: 10.2202/1542-6580.2321

[10] Liu J, Xu Z, Yi C, Haojun F, Shi B. 1-butyl-3-methylimidazolium acetate as an alternative solvent for type I collagen, J. American Leather Chemists Association (JALCA). 2014;675(6):189–196.

[11] Zoltán Baán. Application of ILs in Catalytic Transfer Hydrogenation, PhD Thesis, Department of Organic Chemistry and Technology/ Budapest University of Technology and Economics, Budapest, Hungary, 2008.

[12] Yu Y, Hua L, Zhu W, Shi Y, Cao T, Qiao Y, Hou Z. Ionic liquid–catalyzed internal redox esterification reaction, Synthetic Communications: An Int. J. for Rapid Communication of Synthetic Organic Chemistry. 2013; 43(9):1287–1298, DOI: 10.1080/00397911.2011.632702

[13] Dawkar VV, Jadhav UU, Chougale AD, Govindwar SP. In: Lignin: Properties and Applications in Biotechnology and Bioenergy, Chapter 20, Ryan J. Paterson (ed.), Nova Science, 2012, p. 499–506, ISBN: 978-1-61122-907-3.

[14] Xu A, Wang J, Wang H. Effects of anionic structure and lithium salts addition on the dissolution of cellulose in 1-butyl-3-methylimidazolium-based ionic liquid solvent systems, Green Chem. 2010;12:268–275.

[15] Lethesh KC, Parmentier D, Dehaen W, Binnemans K. Phenolate platform for anion exchange in ionic liquids, RSC Adv. 2012;2:11936–11943, DOI: 10.1039/C2RA22304J.

[16] Hoogerstraete TV, Jamar S, Wellens S, Binnemans K. Determination of halide impurities in ionic liquids by total reflection X-ray fluorescence spectrometry, Anal. Chem. 2014;86: 3931–3938.

[17] Ambika PPS, Chauhan SMS. Chemoselective epoxidation of substrate containing both electron rich and electron deficient olefins catalyzed by meso-tetraarylporphyrin iron(III) chlorides in imidazolium ionic liquids, Electronic Supplementary Material

(ESI) for New Journal of Chemistry, 2011. http://www.rsc.org/suppdata/nj/c1/c1nj20739c/c1nj20739c.pdf

[18] Yang Y, Wang LB, Zhang Z, Li CM, Fu XL, Gao GH., [Bmim]OAc catalyzed Michael addition of active methylene to α, β-unsaturated carboxylic esters, Chem. Res. Chinese Universities. 2010; 26(4):554–557

[19] Clough C, Griffith J, Sulaiman MR, Corbett P, Welton T. Alkylation of 1-methylimidazole with 1-chlorobutane; the ionic liquid 1-butyl-3-methylimidazolium chloride. SCHEMSPIDER. Published Jul 03, 2014, DOI: 10.1039/SP747

[20] Harjani JR, Nara SJ, Salunkhe MM, Sanghvi YS. Transprotection of silyl ethers of nucleosides in $FeCl_3$ based ionic liquids. Nucleosides, Nucleotides, and Nucleic Acids, 2005;24(5–7):819–822.

[21] Corbett PJ. The Synthesis and Utilisation of Ionic Liquids in the Removal of Harmful Impurities from Fuel, PhD Thesis, Imperial College London, 2017.

[22] Löwe H, Axintea RD, Breuch D, Hofmann C, Petersen JH, Pommersheim R, Wang A. Flow chemistry: Imidazole-based ionic liquid syntheses in micro-scale, Chem. Eng. J. 2010;163:429–437

[23] Science lab, "MSDS for Imidazole", available at: http://www.sciencelab.com/msds.php?msdsId=9927195

[24] Science lab, "MSDS for 1-Iodobutane", available at: http://www.sciencelab.com/msds.php?msdsId=9924387

[25] Science lab, "MSDS for Dimethyl carbonate", available at http://www.sciencelab.com/msds.php?msdsId=9923808

[26] Science lab, "MSDS for Acetic acid", available at: http://www.sciencelab.com/msds.php?msdsId=9922769

[27] Linde Gas, "Hydrogen iodide MSDS", available at: http://signupmonkey.ece.ucsb.edu/wiki/images/6/60/Hydrogen_Iodide_MSDS.pdf

[28] Chemical Book, "MSDS for 1-butylimidazole", available at: http://www.chemicalbook.com/ProductMSDSDetailCB6330927_EN.htm

[29] Science lab, "MSDS for 1-methylimidazole", available at: www.sciencelab.com/msds.php?msdsId=9926068

[30] Evonic industries, "GPS Safety Summary for 1-Chlorobutane" available at: https://www.google.ae/search?q=GPS+Safety+Summary%2C+1-Chlorobutane&rlz=1C1AKJH_enAE610AE610&oq=GPS+Safety+Summary%2C+1-Chlorobutane&aqs=chrome..69i57.519j0j7&sourceid=chrome&es_sm=93&ie=UTF-8

[31] Science Lab, "MSDS for Silver acetate", available at: http://www.sciencelab.com/msds.php?msdsId=9927254

[32] Science lab, "MSDS for Silver chloride", available at: http://www.sciencelab.com/msds.php?msdsId=9927255

[33] Harris D. Quantitative Chemical Analysis, 8th ed. Freeman Publishing, 2010, 674–685.

[34] Marzouk SA. http://www.picotech.com/experiments/ph-measurements

[35] High-resolution data acquisition, Pico Technology, Available at: https://www.picotech.com/data-logger/adc-20-adc-24/precision-data-acquisition

[36] Chemical Reactions and Kinetics (purdue.edu)

[37] http://www.ecs.umass.edu/cee/reckhow/courses/572/572bk4/572BK4.html

[38] Kital et al. Journal of Analytical Science and Technology, 2020, 11:41, 13 pages. https://doi.org/10.1186/s40543-020-00238-2

Chapter 2

Fluorinated Ionic Liquids as Task-Specific Materials: An Overview of Current Research

Nicole S.M. Vieira, Margarida L. Ferreira, Paulo J. Castro, João M.M. Araújo and Ana B. Pereiro

Abstract

This chapter is focused on the massive potential and increasing interest on Fluorinated Ionic Liquids (FILs) as task-specific materials. FILs are a specific family of ionic liquids, with fluorine tags equal or longer than four carbon atoms, that share and improve the properties of both traditional ionic liquids and perfluoro surfactants. These compounds have unique properties such as three nanosegregated domains, a great surfactant power, chemical/biological inertness, easy recovery and recyclability, low surface tension, extreme surface activity, high gas solubility, negligible vapour pressure, null flammability, and high thermal stability. These properties allied to the countless possible combinations between cations and anions allow the design and development of FILs with remarkable properties to be used in specific applications. In this review, we highlight not only the unique thermophysical, surfactant and toxicological properties of these fluorinated compounds, but also their application as task-specific materials in many fields of interest, including biomedical applications, as artificial gas carries and drug delivery systems, as well as solvents for separations in engineering processes.

Keywords: fluorinated ionic liquids, task-specific materials, artificial gas carriers, drug delivery systems, separation processes

1. Introduction

Perfluorocarbons (PFCs) consists of a large group of man-made chemicals available worldwide in many different fields since the 1940's [1]. The numerous applications of PFCs in different areas relies on their distinctive physical and chemical characteristics (water and oil repellence, thermal and chemical stability, surfactant behaviour, low polarity, weak intermolecular interactions, and reduced surface tension), [1–3] highly fomented by the fluor-carbon moiety [1–3]. These compounds are widespread in consumers life through plastics, fire retardants, dyes, surfactants, polymers, and pharmaceuticals, among others [1–6]. Benign PFCs have been used in the development of biomedical applications, such as emulsions, [7, 8] imaging agents, [9, 10] biocompatible lubricants, [11] oxygen therapeutics, [12] pulmonary delivery agents, [13] and theranostic agents [14]. On the other hand, perfluoroalkyl acids (PFAs) and fluorinated greenhouse gases (F-gases) belong to a class of persistent chemicals, widely used in industrial and commercial products [1, 2, 5, 6]. Due to their high global warming potential (GWP), long atmospheric lifetime, persistency, and mobility, these compounds have been found in several contaminated sites, [2, 15] including water, soils, biota and food [16–18]. Major concerns about their toxicity and bioaccumulation limit their use and encourages their replacement [1, 2, 5].

In the last decades ionic liquids (ILs) have emerged as new engineering solvents. The application of these compounds has aroused in many different subjects, including catalysis, electrochemistry, extraction and separation processes, pharmaceutical and biomedical applications [19–25]. This massive use of ILs is supported by their unique thermophysical properties and limitlessness combinations between anions and cations [19, 26, 27]. Their title of "green solvents" is corroborated by an almost negligible vapour pressure at room temperature and reduced flammability [19, 26]. Additionally, the increased research about the cytotoxicity and environmental toxicity of these compounds reinforces that their possible harmful behaviour is dependent on the cation-anion tested combination [28]. Due to their complexity and variety, ILs have been categorized in several families according either to their properties or to their applications [29].

This chapter is focused on the use of a less explored ILs family, the fluorinated ionic liquids (FILs), defined as ILs with fluorine tags equal or longer than four carbon atoms [30–33]. The fluorinated tags can create one nanosegregated domain distinct from polar and apolar (hydrogenated) [32, 33]. FILs combine the exceptional properties of conventional ILs (high thermal stability, negligible vapour pressure, reduced flammability, and greener potential) with the greatest properties of traditional PFCs (chemical and biological inertness, reduced surface tension and increased surfactant behaviour). In contrast to the low solubility and toxicity intrinsic to many highly fluorinated compounds, some novel FILs have been designed with completely water miscibility [34, 35] and negligible toxicity, [30, 36, 37] furthering its use in more green engineering processes and biomedical applications. In spite of these outstanding properties, scarce information is available in literature and research is mainly focused on their synthesis and characterization, [38] electrochemical properties, [39] gas solubilities [40] and application as reaction media [38, 41].

This chapter covers the main assets of these FILs, namely their thermophysical and structural properties, aggregation and surfactant behaviour, cytotoxicity, acute ecotoxicity and biodegradation. Additionally, a more detailed approach throughout the application of FILs as task-specific materials in several areas comprise the analysis of a series of works. It is evidenced the progress of FILs either in biomedical applications, or in engineering separation processes.

2. Properties of fluorinated ionic liquids

The characterization of FILs properties and the influence of the different cation/anion combinations on these properties is still critical to head these specific materials to the potential applications. FILs have enhanced properties due to the nanosegregated structuring into three different domains, one polar and two apolar (hydrogenated and fluorinated), making them an alternative solvent with new improved mechanisms of solubilization of different compounds (see **Figure 1**) [31–33]. The manipulation of the nanosegregation behaviour and intra- and intermolecular interactions of FILs allows the control of thermal and thermophysical properties, toxicity, solubility capacity or hydrophobicity of FILs.

In this section, it is emphasized how the formation of the new fluorinated domain and the structural features influence the properties of FILs. The properties

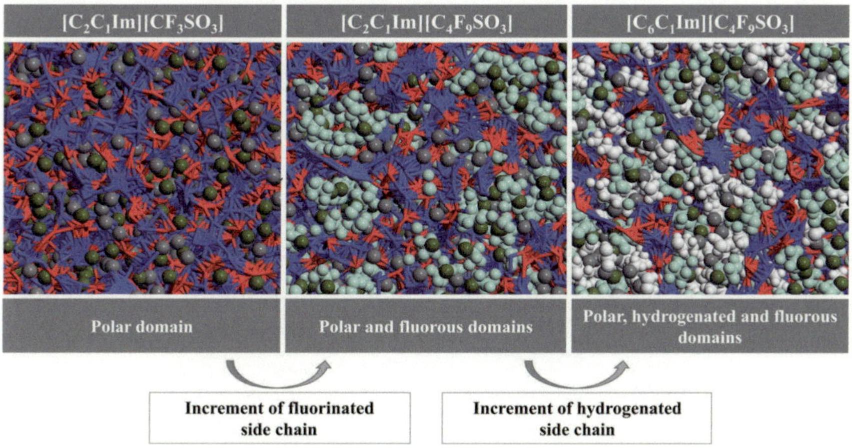

Figure 1.
Formation of three nanosegregated domains of $[C_2C_1Im][CF_3SO_3]$, $[C_2C_1Im][C_4F_9SO_3]$ and $[C_6C_1Im][C_4F_9SO_3]$ FILs. The red and blue sticks represent negative and positive charges, indicating the segregated polar network in the three ILs. The green space-filled areas represent the fluorinated domains. The grey space-filled areas indicate the hydrogenated moieties segregated. Adapted from [42].

of FILs, such as melting point, thermal stability, density, viscosity, refractive index, ionic conductivity and surface tension [30, 33, 42–50] are discussed along with the FILs self-aggregation behaviour in aqueous solutions [34, 35, 50–52]. A close sight on the biocompatibility of FILs by examining their toxicological and biodegradability properties is also included for discussion [30, 36, 37].

2.1 Thermophysical properties

2.1.1 Phase behaviour and thermal properties

The phase behaviour of pure FILs is determined by the melting, solid–solid and glass transitions while the thermal stability is defined by the decomposition temperature. These properties are determinant to define the liquid range of application, allowing a wisely choice of a fluid to a specific task. Several works include the thermal characterization of the FILs depicted in **Table 1** [30, 33, 42–47, 50]. In the case of FILs where the formation of three domains occurs, due to long enough hydrogenated (up to 6 carbons) and fluorinated (up to 4 carbons) chains (**Figure 1**), a rich phase behaviour is found, with a high number of solid–solid transitions. This indicates the ability of FILs domains to rearrange into different structures until the complete melting, proving the high influence of the nanosegregation [33, 46].

The different structural features of FILs can impact the melting and decomposition temperatures, and much work has been done to find trends to design FILs with tuned thermal properties [30, 42, 45, 47, 50]. The melting and decomposition temperatures of several FILs can be found in the **Table 2**. In the case of $[C_nC_1Im][C_4F_9SO_3]$ FILs family, it was found that the increment of the cationic hydrogenated chain increases the melting temperature and decreases the decomposition temperature [42, 47]. The increase of the anionic fluorinated chain also rises the melting point. However, the thermal stability is maintained constant at a considerable high temperature [42, 47]. Moreover, FILs based on $[C_nF_{2n+1}SO_3]^-$ anions have a much higher thermal stability than ILs conjugated with $[C_nF_{2n+1}CO_2]^-$ anions [42, 45, 50]. The type of cation and its functionalization also has a great

Cations structure	
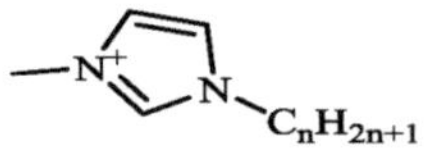 $[C_nC_1Im]^+$ n = 2, 4, 6, 8, 10 and 12 1-Alkyl-3-methylimidazolium	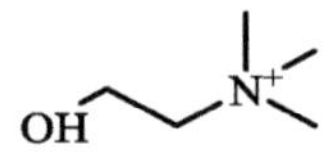 $[N_{1112(OH)}]^+$ (2-Hydroxyethyl)trimethylammonium
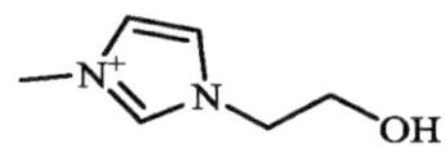 $[C_{2(OH)}C_1Im]^+$ 1-(2-Hydroxyethyl)-3-methylimidazolium	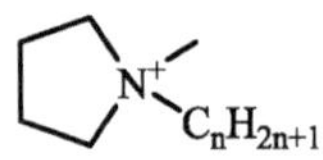 $[C_nC_1pyr]^+$ n = 2 and 4 1-Alkyl-1-methylpyrrolidinium
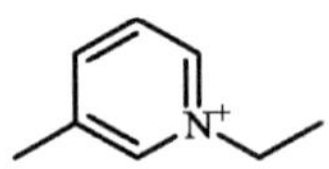 $[C_2C_1py]^+$ 1-Ethyl-3-methylpyridinium	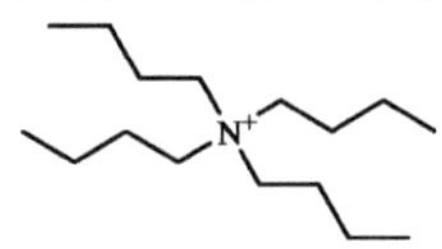 $[N_{4444}]^+$ Tetrabutylammonium
Anions structure	**F-gases**
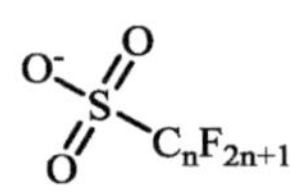 $[C_nF_{2n+1}SO_3]^-$ n = 1, 4 and 8 Perfluoroalkyl sulfonate	 **R-32** Difluoromethane
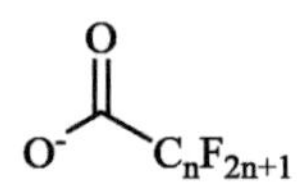 $[C_nF_{2n+1}CO_2]^-$ n = 4 and 8 Perfluoroalkyl carboxylate	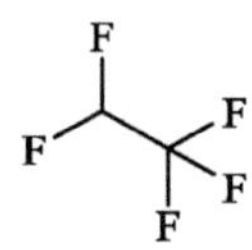 **R-125** Pentafluoroethane
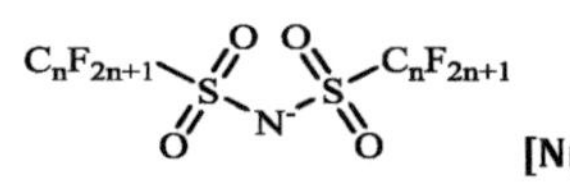 $[N(C_nF_{2n+1}SO_2)_2]^-$, n = 1 and 4 Bis(perfluoroalkylsulfonyl)imide	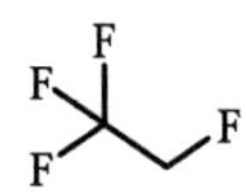 **R-134a** 1,1,1,2-Tetrafluoroethane
Perfluorinated acids	
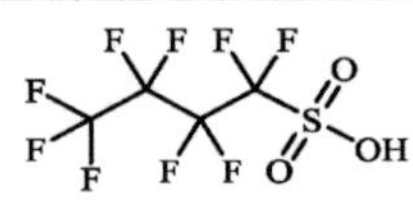 $[C_4F_9SO_3H]$ Nonafluoro-1-butanesulfonic acid	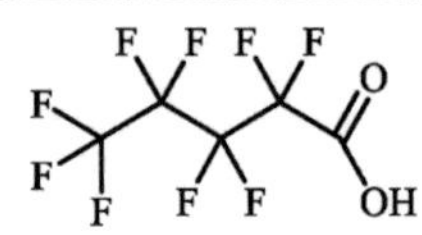 $[C_4F_9CO_2H]$ n-Perfluoropentanoic acid

Table 1.
Structure and nomenclature of the ions constituting the FILs and of the F-gases studied for absorption in FILs and in deep eutectic solvents, prepared with the illustrated perfluorinated acids.

influence in both thermal properties, and a carefully analysis must be performed when choosing a FIL for a specific ending [30, 33, 42, 45, 46, 50].

The FILs based on long fluorinated chains (e. g. $[N(C_4F_9SO_2)_2]^-$) have a very high melting temperature, automatically reducing the liquid operating range. Eutectic mixtures of FILs can be the solution to solve this handicap. The evaluation of the solid–liquid phase behaviour of binary mixtures of FILs showed a high decline of the melting temperature to values close or below room temperature [44]. This does not only increase the liquid range of FILs, but also expands the tuneability of neat FILs.

	T_m K	T_{onset} K	ρ g·cm^{-3}	η m·Pas^{-1}	γ mN·m^{-1}
			$[C_nC_1Im][C_4F_9SO_3]$		
n = 2	293 [42]	627 [42]	1.547 [42]	163.0 [42]	25.14 [43]*
n = 4	286 [47]	638 [47]	1.460 [47]	307.3 [47]	22.83 [43]*
n = 6	297 [30]	627 [30]	1.392 [30]	401.7 [30]	21.36 [43]*
n = 8	308 [30]	621 [30]	1.338 [30]	374.6 [30]	20.57 [43]*
n = 10	307 [47]	627 [47]	1.310 [47]	597.1 [47]	22.05 [43]*
n = 12	311 [42]*	617 [42]*	1.247 [42]*	280.9 [42]*	23.42 [43]*
			$[C_4F_9SO_3]^-$		
$[C_2C_1py]^+$	278 [30]	629 [30]	1.515 [30]	201.8 [30]	26.35 [45]
$[N_{4444}]^+$	327 [30]	587 [30]	1.234 [30]	15319 [30]	22.77 [45]**
$[C_4C_1pyr]^+$	364 [46]	632 [46]			
$[N_{1112(OH)}]^+$	436 [45]	609 [45]			
$[C_{2(OH)}C_1Im]^+$	251 [50]	559 [50]	1.620 [50]	831.6 [50]	
			$[C_4F_9CO_2]^-$		
$[C_2C_1Im]^+$	278 [42]	392 [42]	1.487 [42]	107.5 [42]	
$[C_8C_1Im]^+$	297 [42]	399 [42]	1.292 [42]	307.9 [42]	
$[C_{2(OH)}C_1Im]^+$	295 [50]	433 [50]	1.541 [50]	712.8 [50]	
$[C_2C_1py]^+$	275 [45]	392 [45]	1.454 [45]	147.1 [45]	26.83 [45]
			$[C_8F_{17}SO_3]^-$		
$[N_{4444}]^+$	255 [30]	385 [30]	1.317 [30]	6690 [30]	21.98 [45]
$[C_2C_1Im]^+$	368 [42]	616 [42]			
			$[N(C_4F_9SO_2)_2]^-$		
$[C_2C_1pyr]^+$	428 [45]	619 [45]			
$[C_4C_1pyr]^+$	371 [45]	639 [45]			
$[N_{1112(OH)}]^+$	303 [45]*	622 [45]*	1.674 [45]*	947.1 [45]*	25.04 [45]*

*Experimental data obtained * at 313.15 K and ** at 333.15 K.*

Table 2.
Thermophysical and thermodynamic properties of fluorinated ionic liquids at 298.15 K and atmospheric pressure: melting temperature, T_m; decomposition temperature, T_{onset}; density, ρ; viscosity, η; and surface tension, γ.

2.1.2 Density, transport properties, free volume, and surface tension

Density, transport, free volume, and surface tension properties have high relevance in the biomedical field as well as in the separation and extraction processes for industrial proposes [30, 53]. The structural features of FILs can determine their density, [30, 42, 45, 47, 50] as can be seen in **Table 2**. While the increment of the fluorinated chains increases FILs density, [30, 42, 45] the opposite behaviour is found for the increment of hydrogenated side chain [30, 42, 45, 47]. The carboxylate anions show a lower density comparing with the sulfonate anions [30, 45, 50]. The functionalization of imidazolium cation with a hydroxyl group has shown an increment on density [50]. The cation nature widely affects the density, and each family must be analysed case by case to infer on the applicability of each FIL [30, 42, 45].

The characterization of FILs viscosity, and consequently of their fluidity, was studied in several works, [30, 42, 45, 47, 50] and some of the results can be found in **Table 2**. The results indicate that FILs with longer aliphatic and fluorinated chains increase the viscosity [30, 42, 45, 47]. The FILs composed by $[C_nF_{2n+1}SO_3]^-$ anions also present high viscosity comparing with the $[C_nF_{2n+1}CO_2]^-$ anions [30, 42, 45]. The nature of the FIL cation affects tremendously the viscosity. In the case of bulkier cations, a lower fluidity is found [30, 42, 45]. The addition of a hydroxyl group in imidazolium cations increases the cohesive forces resulting in more viscous fluids [50].

The ionic conductivity has great importance, especially when correlating the molar conductivity with the fluidity obtaining the ionicity of FILs [30, 42]. The ionicity is evaluated by the Walden plot where FILs are classified depending on the distance to an ideal electrolyte [54]. From the ionicity can result information on the formation of aggregates between ions due to low mobility [54]. The analysis of the results shows that the increment of the cationic aliphatic and of the anionic fluorinated chains decrease the ionicity, diverging from the ideal behaviour [30, 42, 45, 47].

The free volume has a high relevance to FILs suitability as enhanced solvents of gases or other compounds with low molecular weight [55]. The relation between refractive index and density allows the calculation of molar free volume effects, evaluating the available space for dissolution of gases [30, 42, 45, 47, 50]. Therefore, the increase of both hydrogenated and fluorinated chain and bulkier cations rise the molar free volume values [30, 42, 45, 47, 50].

The surface tension of FILs is the property that most differs from the conventional ILs, in which the cation's nature has a predominant influence on this property [43, 45, 56]. The values of surface tension for some FILs can be found in **Table 2**. The surface tension of $[C_nC_1Im][C_4F_9SO_3]$ family showed the lowest values existing in the overall ILs literature [43]. The increment of the hydrogenated chain decreases the surface tension up to the lowest value, found for the $[C_8C_1Im]$ $[C_4F_9SO_3]$. The further increase of FILs aliphatic chain resulted in higher values of surface tension, revealing a global behaviour marked by a bowl-shaped trend [43]. The addition of a fluorinated domain in FILs induces a competition with the aliphatic domain to protrude the interface, which dramatically changes the values of surface tension [43]. As long as the hydrogenated chain increases to $[C_8C_1Im]^+$, a rearrangement in the organization between the non-polar domains happens, allowing both to protrude through the top layer. After $[C_8C_1Im]^+$, the aliphatic chain is much larger than the fluorinated chain, and occupies more space at the interface, increasing the values of surface tension [43]. In the case of quaternary ammonium-based FILs it was shown that they have lower values of surface tension comparing with pyridinium cation. In FILs based on ammonium, the increment of the fluorinated chain deeply decreases the surface tension [45].

The FILs properties can be tuned by choosing the cation, anion, length of side chains and functionalization of cation, increasing the possibilities of designing the best task-material. The complete determination of these properties is a complex assignment, requiring a lot of costs and time. To ease this task, theoretical models can be applied to predict their characteristics. An effort has been done in this direction obtaining several models that accurately reproduces the FILs properties of the neat FILs and of the mixtures with gases and aqueous solutions [47–50].

2.2 Aggregation and surfactant behaviour

The behaviour of FILs in aqueous solutions is enhanced in comparison with the PFCs and conventional ILs [34, 35, 50–52]. The selection of nontoxic FILs based on imidazolium, pyridinium (with short aliphatic chains) and cholinium cations conjugated with the $[C_4F_9SO_3]^-$ anion were used to study the self-aggregation behaviour. These compounds are completely miscible in water at all range of concentrations studied in the conductivity profile [34]. The same behaviour was later found for imidazolium-FILs functionalised with a hydroxyl group [50] and some examples are represented in **Figure 2a**. The Liquid + Liquid equilibria of binary systems FIL + water was also analysed to study the solubility of water [35, 52]. The increment of the aliphatic chain in $[C_nC_1Im][C_4F_9SO_3]$ family increases the solubility of water in the FIL-rich phase [35, 52].

The water-rich region was selected to determine the critical aggregation concentrations (CACs) of several FILs [34, 35, 50, 52]. $[C_2C_1Im][C_4F_9SO_3]$ showed three different transitions related to the formation of distinct aggregates. These aggregates were evaluated and associated to different self-assembled structures [34]. These stable self-assembled structures can be the greatest contribution to the full miscibility of FILs in water. **Figure 2b** represents the values of the first CAC, so-called critical micelle concentration (CMC) of FILs [34, 35, 50, 52] and conventional surfactants [57–59]. All the FILs show much lower CMC and FILs with only four carbon atoms have greater aggregation power than the conventional surfactants with eight carbon atoms. The increment of the hydrogenated chain in the $[C_nC_1Im][C_4F_9SO_3]$ family decreases the CMC value, promoting the formation of more, bulkier and better packed structures [35, 52]. The longer fluorinated chains also decrease the CMC values. However, the growth of both nonpolar chains hinders the solubility in water [34, 35, 52]. The pyridinium and tetrabutylammonium cations show slightly lower CMC values comparing with imidazolium, cholinium or pyrrolidinium cations [19, 20, 22].

The FILs behaviour in water was also inferred in the FIL-rich phase by investigating the hydrogen-bonding ability and polarizability through Kamlet-Taft parameters [51]. The results indicate that increasing the fluorinated chain restricts the impact of adding water into ILs, keeping the hydrogen bond acceptance ability constant. This result indicates that the rich aggregation of FILs promotes the aggregation of water in a bulky polar network. The water aggregates expand and drive to the proximity of the polar nanosegregated domains of the FILs due to the higher repulsion of the fluorinated counterparts [51].

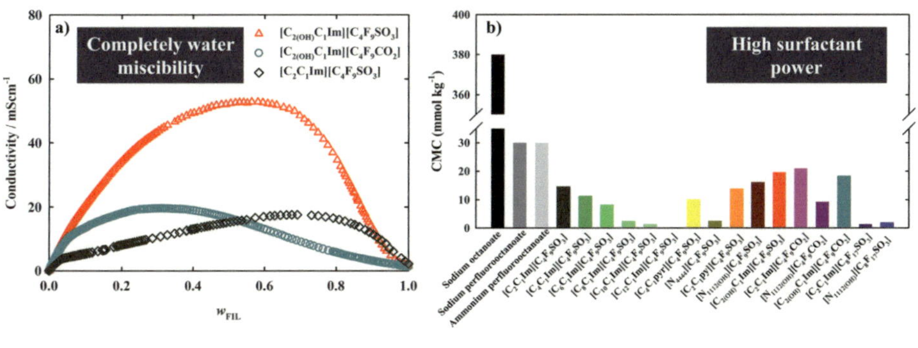

Figure 2.
(a) Complete conductivity profile of FILs in water at 298.15 K and (b) the values of critical micellar concentrations of PFCs (grey bars) and hydrogenated (black bar) surfactants [57–59] and of the FILs (coloured bars) [34, 35, 50, 52].

2.3 Cytotoxicity, ecotoxicity and biodegradation

Cytotoxicity, partition properties, acute ecotoxicity and biodegradation are key parameters to assess the health and environmental risks of these FILs. Knowledge about structure-toxicity relationships is of great interest for the design of biocompatible and greener FILs. The design of these new compounds aims to surpass the persistency, bioaccumulation, and toxicity drawbacks of PFCs [1, 2, 5, 6].

This section provides a critical review of the cytotoxicity in different human cell lines: human colon carcinoma cells (Caco-2), human hepatocellular carcinoma cells (HepG2), human umbilical vein cell line (EA.hy926), and spontaneously immortalized human keratinocyte cell line (HaCaT), representing the risks associated to different routes of biomedical administration [30, 37]. Cytotoxicity screenings, with 4 h [30] and 24 h [37] exposure, were performed in these cell lines. For short-chain based-FILs, such as $[C_2C_1Im][C_4F_9SO_3]$ and $[C_2C_1py][C_4F_9SO_3]$, the overall reduced toxicity can be justified by their high hydrophilicity and surfactant performance [30, 34, 35, 37, 52]. In HaCaT cells, higher EC_{50} values were obtained for both FILs mentioned before and these results can be associated to the intrinsic properties of this cell line [37]. A higher biocompatibility was attained with the cholinium cation conjugated with the $[C_4F_9SO_3]^-$ anion, due to the non-aromaticity and symmetry of this cation, which is also an essential nutrient for cell growth [25, 37, 60]. A similar behaviour was reported for several cholinium alkanoates [61, 62]. The non-aromatic and symmetric $[N_{4444}]^+$ as well as the alicyclic pyrrolidinium cations, conjugated with the $[C_4F_9SO_3]^-$ anion, maintain the cellular viability in Caco-2, HepG2 and EA.hy926 cells [30, 37]. The elongation of the imidazolium hydrogenated alkyl chain length from $[C_2C_1Im]^+$ up to $[C_{12}C_1Im]^+$ prompts the decrease of the cellular viability in the Caco-2 cell line, as depicted in **Figure 3a** [37]. This effect on cellular viability can be due to the presence of delocalized charges or due to the increment of lipophilicity which enhance the disruption of the cell wall [37, 63]. A more pronounced decay on the cellular viability is observed with the increment of the anionic fluorinated side chain length [30, 37]. This effect was noticed for the variation of $[C_4F_9SO_3]^-$ to $[C_8F_{17}SO_3]^-$ or $[N(C_4F_9SO_2)_2]^-$ anions, combined with imidazolium, cholinium and ammonium-based cations [30, 37]. The fluorinated elongation on carboxylate-based anions also engenders a significant reduction of the cellular viability in different cell lines [62]. The increment of the fluorinated domain also enhances the FILs lipophilicity and the charges delocalization, which is traduced in a higher permeation of the cell membranes [37, 64]. Inside the cell compartment, free fluoride ions are formed by

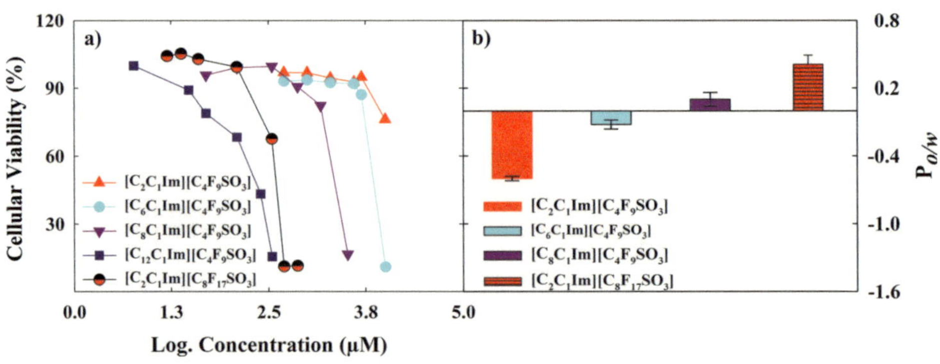

Figure 3.
(a) Cellular viability for imidazolium-based FILs with the increment of hydrogenated and fluorinated alkyl side chain length; (b) Effect of the hydrogenated and fluorinated alkyl side chain length on the 1-octanol/water partition coefficient ($P_{o/w}$) of imidazolium based FILs. Adapted from [37].

hydrolytic cleavage, which can interfere with the cellular mechanisms leading to cell death [37, 64].

The increment of the lipophilicity as result of the elongation of both hydrogenated and fluorinated alkyl side chain was confirmed through the 1-Octanol/water partition coefficients ($P_{o/w}$) of different FILs [37]. As depicted in **Figure 3b**, the $P_{o/w}$ increases with the increment of the hydrogenated side chain length from $[C_2C_1Im]^+$ to $[C_8C_1Im]^+$ [37]. This increment is associated to a greater lipophilic behaviour, caused by stronger van der Waals interactions between the FIL alkyl side chain and the hydrophobic region of the organic solvent, promoting their solubility in the organic media [37, 65]. This elongation also decreases the polarity and the acidity of these compounds, and consequently their interaction with water media [37, 65]. The increment on the anion core from $[C_4F_9SO_3]^-$ to $[C_8F_{17}SO_3]^-$ has a more pronounced effect in the partition properties, as illustrated in **Figure 3b** [37]. These results were associated to an enhanced solubility in lipophilic solvents endorsed by the fluorinated moiety [37, 66]. Finally, the partition properties of both $[C_2C_1Im][C_4F_9SO_3]$ and $[C_2C_1py][C_4F_9SO_3]$ are quite similar due to the highly acidic methylene groups in the constitutive rings [65]. Nevertheless, the partition properties of the studied FILs indicate that they not accumulate or concentrate in the environment [37].

An environmental hazard assessment is also essential in the context of sustainability and green chemistry. An ecotoxicological screening to evaluate the impact of FILs in aquatic environment was performed in marine bacterium *Vibrio fischeri*, crustacean *Daphnia magna*, and in *Lemna minor* plant [36]. This screening was made in aquatic species owing to the selected FILs unique water miscibility [34, 35]. Briefly, all tested FILs present a reduced ecotoxicity for the mentioned species [36]. The EC_{50} values indicate that FILs based on the imidazolium cation conjugated with $[C_4F_9SO_3]^-$ anion are more toxic than FILs based on other cations conjugated with the same anion [36]. The $[C_4F_9CO_2]^-$ anion is also less toxic than the sulfonate equivalent, except for the hydroxylated based imidazolium cations in *Daphnia magna and Lemna minor* [36]. Even so, the $[C_4F_9SO_3]^-$ based anion are less toxic than the bis(trifluoromethylsulfonyl)imide ($[N(CF_3SO_2)_2]^-$) anion for both *Vibrio fischeri* and *Daphnia magna* [36, 60]. Furthermore, both cholinium and hydroxylated imidazolium cations are the least toxic in the three aquatic species [61]. The functionalization of the imidazolium cation decreases the lipophilicity of these compounds, and consequently decreases their overall toxicity [36]. Finally, it must be stated that based on *Daphnia magna* and *Lemna minor* EC_{50} values and accordingly to the "Globally Harmonized System of Classification and Labelling of Chemicals", these FILs do not need to be categorized in terms of acute aquatic hazard [36]. It must be noticed that both cytotoxicity and ecotoxicity results are highly dependent on the target organisms and exposure times, [36, 37, 62] then different species and long-term effects of these compounds must be accessed prior to a large-scale application.

The microbial degradation of some FILs showed that short chain-based imidazolium FILs are highly resistant to biodegradation, even with the incorporation of hydroxyl groups. A certain biodegradability occurred in the short chained pyridinium-based FIL, associated to the oxidation of the alkyl side chain [36, 67, 68]. However, some variability is associated to the biodegradation of these cation that must be associated to the differences in microbial compositions involved in the degradation process [67, 68]. The higher degrees of biodegradation obtained with the cholinium-based FILs is only related to the cation core degradation that retains 75% of the oxidizable carbon [36]. To overcome the highly resistance associated to these compounds, removal or degradation alternative routes must be studied. According to these published results a proper combination between cations and short chained fluorinated anions may result in biocompatible FILs with potential to be biodegradable by alternative routes. These biocompatible FILs can support the fields of FILs as task-specific materials in a broad range of fields, from biomedical to reaction media in industrial processes.

3. Applications of fluorinated ionic liquids

3.1 Biomedical applications

3.1.1 Artificial gas carriers

The need of new products to replace the blood transfusions appeared in the beginning of the 21th century as a consequence of cross-infections derived from the human immunodeficiency virus (HIV) [4, 69]. The lack of safety and trust allied with the severe shortages and increased demand of blood supplies have contributed to the search of an ideal artificial gas carrier (AGC) [4, 69]. PFCs-based emulsions are among the substances under clinical trials used to substitute the red blood cells in critical situations such as acute blood loss [4]. However, the PFCs have several handicaps that can restrict their usage as AGCs, such as high vapour pressures and poor solubility in water. With the aim to solve these limitations, FILs appeared as a solution to replace the PFCs fully or partially in AGC emulsions. Different works have been developed to infer on this prospect [34, 35, 49, 50, 70, 71]. The results show the possibility to design FILs with complete water miscibility, which solves one of the greatest handicaps [34, 35, 50]. The study of phase equilibria between FILs and two PFCs, perfluorodecalin and perfluorooctane, indicated that the enthalpic contributions are larger than the entropic contributions, which results in a favourable process of solvation of PFCs by FILs [70]. The high surfactant behaviour of FILs is also a huge advantage because it enables the stabilization of AGC emulsions, which can be favourable to reduce the usage of excipients and to enhance the solubilization of the respiratory gases [34, 35, 50]. The reduced cytotoxicity and ecotoxicity determined for FILs with the characteristics above mentioned strengths the possible use of these compounds in the biomedical field [30, 36, 37]. The greatest aspect that spurs the use of FILs as potential substitutes of PFCs in AGC emulsions is their higher ability to solubilize oxygen, carbon dioxide and nitrogen, compared to the conventional fluorine-containing ILs and with PFCs [49, 71]. However, the formulation of an emulsion with high efficacy and the implementation of tests on the physiological safety and other health studies must be carried out before applying FILs.

3.1.2 Drug delivery systems

Although there are several studies dealing with ILs for the solubilisation and stabilization of proteins, [23, 72] dissolution of low soluble active pharmaceutical ingredients (APIs), [23, 24] and development of drug formulations and delivery systems, [23–25, 73] the application of FILs in this field of pharmaceutical development is quite unexplored. Our research group initiated a pioneering research line to use FILs as drug delivery systems (DDSs) [74–76]. These novel biocompatible carriers can overcome the problems associated to proteins administration (e.g. sensibility to environmental conditions, short-half lives in blood stream, structural conformation and hydrophobic/hydrophilic nature that hamper the *in vivo* delivery) [77, 78] and their traditional delivery platforms (low stability, uncontrolled release, and low encapsulation efficiency) [79]. FIL-based DDSs have been shown the potential to increase the safety and effectiveness of the therapeutic biomolecules, reducing the dosage needed and enabling a time and site-specific release [74–76].

The application of FILs as DDSs and stabilizing agents was firstly evaluated for two different model proteins, lysozyme, and bovine serum albumin (BSA) [74, 75]. Lysozyme is a protein with antiviral, antitumor and immunological properties, [80] whereas BSA is involved in organism homeostasis and in the transport of several components essential for several vertebrates' body functioning [81]. For these applications, FILs based on imidazolium, pyridinium and cholinium cations, conjugated with $[C_4F_9SO_3]^-$ and $[C_4F_9CO_2]^-$ anions were selected due to biocompatibility and improved surfactant behaviour [30, 31, 34–37]. The tested FILs concentrations cover values above and below their CMCs values (**Figure 2b**) [34, 52]. Concentrations above CMC were chosen due to their ability for self-assembling in micellar structures that can be used to protect, encapsulate, and deliver the therapeutic proteins [34, 52]. The stability of both proteins in the presence of FILs was determined based on the variations observed in the melting temperature of the biomolecules [74, 75]. The stability of lysozyme is not significantly affected by the incorporation of FILs, and only a slight decrease was achieved with $[C_2C_1py][C_4F_9CO_2]$ with a minor reduction of 2% in the melting temperature of the protein [74]. However, for BSA the melting temperature increases for all tested FILs concentrations, suggesting a stabilization of the protein [75]. These distinct results indicate a specific interaction between FILs and each tested protein [74, 75]. The differences among the interactions of the two biomolecules with FILs were also supported by structural studies. Both circular dichroism (CD) and fourier transformed infrared spectroscopy results suggest no substantial lysozyme structural modifications in the presence of cholinium and $[C_2C_1Im][C_4F_9SO_3]$ FILs, respectively [74]. For BSA, a slight increment on molar ellipticity and α helical content, followed by a β sheet and random coil reduction, observed in CD results, indicate a stabilization of the secondary structure, and a more compact state of the protein with $[N_{1112(OH)}][C_4F_9SO_3]$ [75, 82]. Furthermore, in the presence of FILs, the biological activity of lysozyme increased, even at concentrations where the encapsulation of the protein inside the micelles occurs [74]. Although there are differences in the interactions between the two different proteins and the FILs, the stability, activity and secondary structure of biomolecules are not negatively impacted by the selected fluorinated compounds [74, 75].

The aggregation behaviour of different FILs was analysed in the protein medium. No significant variations were achieved in the FILs self-aggregation process in aqueous solutions [34, 74, 75]. To prove the encapsulation of lysozyme in the aggregates of FILs, the self-assembled structures were studied through dynamic light scattering (DLS) [74]. As illustrated in **Figure 4a**, an encapsulation of the protein at a concentration approximately twice the FILs CMC (1.2% v/v) is expected based on the disappearance of the intensity peak of lysozyme (~ 4 nm) [74]. This encapsulation is driven by the fluorinated surfactant core of the FILs since the lysozyme characteristic peak remains present for the non-surfactant ILs [74]. This encapsulation was indorsed spectrophotometrically with the concentration of lysozyme in solution being reduced with the addition of 1.2% v/v $[C_2C_1Im][C_4F_9SO_3]$ [74]. Moreover, the FIL-protein aggregates became more stable after 24 h and a maximum stabilization was verified after 96 h [74]. The lysozyme encapsulation in $[C_2C_1Im][C_4F_9SO_3]$ was also evidenced, illustrated in **Figure 4b** and **c** [74]. **Figure 4b** depicts the solution of lysozyme with 1.2% v/v of $[C_2C_1Im][C_4F_9SO_3]$ analysed by transmission electron microscopy (TEM), where an external darker counter surrounding the aggregates of FILs is associated to the heavier elements present in the anion, in contrast to the lighter grey shades of the lysozyme

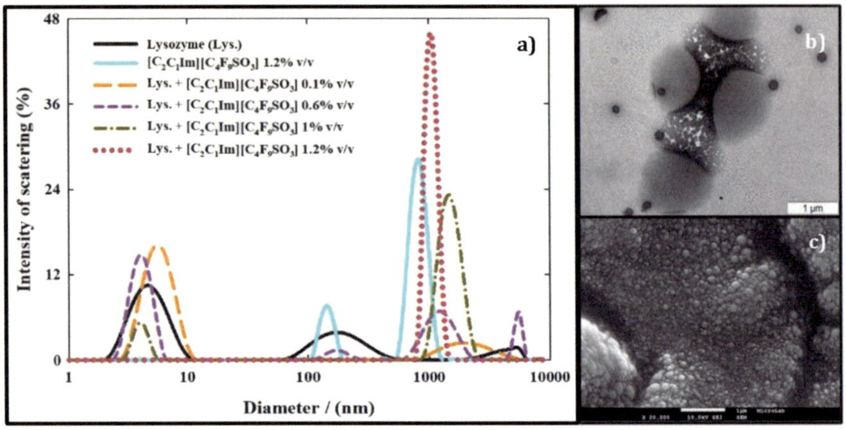

Figure 4.
(a) DLS spectra of lysozyme in buffered medium upon the addition of $[C_2C_1Im][C_4F_9SO_3]$ at several concentrations; (b) TEM image of $[C_2C_1Im][C_4F_9SO_3]$ 1.2% v/v with lysozyme; (c) SEM image of $[C_2C_1Im][C_4F_9SO_3]$ 1.2% v/v with lysozyme. Adapted from [74].

[74]. Moreover, the micellar sizes obtained by TEM are similar to the hydration diameters measured by DLS. A qualitative analysis through scanning electron microscopy (SEM), **Figure 4c**, reveals an external surface of the solution containing lysozyme with 1.2% v/v of $[C_2C_1Im][C_4F_9SO_3]$ similar to the FILs blank solution depicted in [74].

The interaction and the encapsulation between $[C_2C_1Im][C_4F_9SO_3]$ and BSA was proved through isothermal titration calorimetry (ITC) [75]. BSA interacts with the $[C_2C_1Im][C_4F_9SO_3]$ monomers causing conformational changes, as well as hydrogen bonding and hydrophobic interactions [75]. The aggregation of $[C_2C_1Im][C_4F_9SO_3]$ in buffer determined by conductimetry was also supported by the ITC measurements. However, ITC indicates that the interaction between BSA and FIL is stronger than the FIL self-aggregation [75]. A different interaction between BSA and the FIL aggregates, not identified in the conductivity measurements, strongly supports the encapsulation of this protein inside the FILs aggregates [75].

After the first proof of concept dealing with the encapsulation of lysozyme inside the FIL aggregates, the optimal incubation temperature of the protein during 24 h was determined at 4 °C without a significant loss of protein activity [76]. The encapsulation efficiencies of lysozyme in both $[C_2C_1Im][C_4F_9SO_3]$ and $[C_2C_1py][C_4F_9SO_3]$ at 1.8% v/v (3 times higher than CMC) range from 69.4 to 83.4%, values similar or higher than the obtained with other traditional platforms [76]. This lysozyme remains encapsulated up to 12 h post-incubation at 4 °C, without significant losses of biological activity [76]. This longer retention of the biomolecule inside the FILs aggregates can be caused by the high stability of the fluorinated counterpart of the IL, as well as by the interaction between FIL and protein [76]. Furthermore, the biomolecule release was accomplished after the application of several external stimuli [76]. With the increment of temperature up to 37 °C, simulating the average body temperature, lysozyme is completely released from the aggregated structures after 6 h [76]. This complete release was also achieved after the exposure to an ultrasound bath with a frequency of 80 kHz during 1 h [76]. This approach can be applied for a site specific and controlled delivery of therapeutic proteins through FILs based DDS. Furthermore, within the same time frame at 42 °C the protein released range from 57% and 39% to $[C_2C_1Im][C_4F_9SO_3]$ and $[C_2C_1py][C_4F_9SO_3]$ based DDS, respectively, suggesting that under a pathological condition the protein can be released at some relevant extent after 1 h post administration [76]. The biological activity of the released protein remains above 50% for all the tested scenarios, except for the release after 12 h at 37 °C [76]. Then, biocompatible FILs can be designed to encapsulate different therapeutic proteins with good levels off encapsulation efficiencies promoting a site specific and thermo responsive release under different external stimuli. The differences among the effect of FILs in both lysozyme and BSA support the need to further study the interactions of these fluorinated compounds with other therapeutic biomolecules prior the design of the DDS.

3.2 Separation of fluorinated greenhouse gases using FILs

Currently, there is a great interest in the development of technologies to reduce the emissions of greenhouse gases (GHGs) into the atmosphere. F-gases, including hydrofluorocarbons (HFCs), PFCs, and sulphur hexafluoride (SF_6), are major contributors to GWP with long atmospheric lifetime. The most predominant F-gases used in refrigeration include 1,1,1,2-tetrafluoroethane (R-134a) and difluoromethane (R-32), alone or in blends with other F-gases, such as pentafluoroethane (R-125). In order to accomplish the international goals to reduce the emissions of GHGs, new refrigerants with lower GWP are being investigated and great research efforts are being made aiming to develop technologies to selective separate value-added F-gases from depleted refrigerants. These technologies lead to a reduction of gas emissions and promote the use of recycled F-gases. However, the separation of F-gases faces a major challenge, particularly in the cases of gas blends with an azeotropic or near-azeotropic behaviour. R-410A is widely used in the refrigeration sector but has a high GWP. Therefore, this refrigerant is one of the focus of the EU HFC phase-down [83]. This blend is a near-azeotropic system of R-32 and R-125 and therefore the separation of its individual components is hampered [83]. Consequently, there is a growing interest in the search for new efficient, low-energy, and sustainable separation processes.

The solubilization of F-gases in FILs is a poorly explored area. Most work has been done with imidazolium-based ILs composed of the $[N(CF_3SO_2)_2]^-$, tetrafluoroborate ($[BF_4]^-$) or hexafluorophosphate ($[PF_6]^-$) anions for the solubilization of different HFCs [84–87]. Gas solubility in ILs is an interplay of different phenomena with: (i) the enthalpic contribution of the intermolecular interactions between gas molecules and the absorbent and; (ii) the entropic contribution of the accommodation of gas molecules in the cavities of the absorbent. A positive correlation is found between the degree of fluorination of the ILs and the solubilization of HFCs [87, 88]. Additionally, the fluorination of the cation was shown to play a major role in the solubilization of PFCs [89] and HFCs [90] in 1-alkyl-3-methylimidazolium based ILs. The structures and the fluorination degree of the gases also strongly affect their solubilization into ILs. Solubilities of a variety of F-gases in $[C_2C_1Im][N(CF_3SO_2)_2]$ have been evaluated experimentally, and by modeling with soft-SAFT equation. These studies demonstrated the importance of the establishment of hydrogen bonds between the gas molecules and the absorbent. Both entropic effects, resulting from higher chain length/volume, and enthalpic effects, resulting from higher dipole moment, are suggested to increase gas solubility [91].

FILs present particular properties that distinguish them from mere fluoro-containing ILs, such as the ones with the $[N(CF_3SO_2)_2]^-$, $[BF_4]^-$, and $[PF_6]^-$ anions. Their ability to form three nanosegregated domains with different behaviours and the existence of countless cation/anion combinations increase the range of possible interactions (van der Waals, coulombic, and hydrogen bonding), making them ideal three-in-one solvent for the separation of F-gases [33].

When evaluating the absorption capacities of traditional ILs and of FILs for the selective capture of R-32 (**Table 1**), a positive relation between the fluorination degree of the anion and the solubilization of this gas was reported [91]. This behaviour is similar to what is observed when the size of the hydrogenated alkyl chain in the cation of fluoro-containing imidazolium-based ILs increases, [86, 88, 92, 93] and can be explained by the entropic contribution of the accommodation of gas molecules in the cavities of absorbents with higher molar volume. Moreover, when the absorption of R-125 and R-134a in the abovementioned ILs was studied, a higher solubility capacity of FILs in comparison to mere fluoro-containing ILs was observed [83]. This demonstrates the relevance of the FILs nanosegregated domains for gas solubility, either by increasing the free volume for the accommodation of gas molecules or by increasing the number of possible gas-absorbent interactions. Lower solubilities have been obtained in mere fluoro-containing and in FILs to R-125 in comparison to R-134a. This has been explained by the decrease in the number of interactions with the absorbent as a consequence of the reduced number of hydrogen atoms in R-125, [89] or by a decrease in the flexibility of R-125, as consequence of a higher number of fluorine atoms [91]. By playing with the different factors involved in the solubilization of F-gases in ILs, namely the constitution of the cations and anions of the IL, temperature, pressure and others, it is possible to develop processes where the solubilization of one gas is favored in relation to other gas, or gases, present in the same mixture [91]. In this way, while the separation of the binary mixtures R-134a + R-125 and R-32 + R-125 was demonstrated to be improved using fluoro-containing ILs, lacking an alkyl fluorinated chain, the separation of the mixture R-134a + R-32 might be improved by utilizing FILs.

The increased solubility of F-gases in FILs supports the use of these absorbent as an alternative to conventional ILs with longer hydrogenated chains, which present higher toxicity [83]. Other study focused on evaluating the viability and costs of an absorption technology in near-industrial conditions for the capture of R-32 and R-134a (with HFC recoveries above 90%) from a dilute gas stream, using FILs or mere fluoro-containing ILs as absorbents. In this study a COSMO-based/Aspen Plus methodology was applied to evaluate the influence of ILs structure, HFC partial pressure, operating temperature, and FIL/IL mass flow on the recovery of HFCs [94].

The development of separation processes based on ILs may face some obstacles due to the unfavorable properties of some of these compounds, such as the toxicity of those with long fluorinated alkyl side chains, poor biodegradability, high viscosity, high-cost production, and high melting temperature. As aforementioned, the solubility of F-gases is favored when the number of fluorine atoms in ILs is increased, but this is also associated with higher melting temperature and to a decrease in the range of temperatures in which FILs can be operated at the liquid state. In this sense, deep eutectic solvents (DESs) are emerging as a versatile alternative to ILs, with low vapour pressure, nonflammability, high tuneability, and improved properties for application at process level. DESs are systems in which the charge delocalization occurring through hydrogen bonding between a hydrogen bond acceptor (HBA) and a hydrogen bond donor (HBD) is responsible for decreasing the melting point of the mixture relatively to the individual components. Experimental studies regarding the solubility of refrigerants in DESs are scarce [95–98]. The solubility of R-134a in DES prepared by combining the IL $[C_2C_1Im][Cl]$ as hydrogen-bond acceptor (HBA) and 4-carbon perfluoroalkyl acids as hydrogen-bond donors (HBDs), was studied using both experimental solubility data and a theoretical model based on the soft-SAFT equation of state [89]. Additionally, the solubilization of F-gases was studied in DESs prepared by mixing high melting temperature FILs with perfluoropentanoic acid or nonafluoro-1-butanesulfonic acid (**Table 1**) [99]. The selected FILs were composed of different cations (cholinium, imidazolium, or a tetrabutylammonium cation) and anions with 4-carbon or

8-carbon perfluoroalkyl chains (**Table 1**). The melting temperatures of the prepared eutectic mixtures were significantly lower than the one of the neat FILs, which allowed to take advantage of the properties of FILs for the selective separation of F-gases, in a wider liquid range for F-gases solubilization [99].

4. Conclusions

In this chapter, the application of FILs as task-specific materials was fully described to be employed in both biomedical and engineering separation processes. The characteristic fluorinated domain and the different ions structural features prove to have a dominant effect on thermophysical and thermodynamic properties of FILs. Moreover, FILs have great surfactant behaviour and complete miscibility in water systems. The design of biocompatible and eco-friendly FILs without comprimising their surfactant behaviour was demonstrated which ultimate the applicability of FILs as enhanced materials comparing with PFCs and conventional fluorinated ILs.

The applicability of biocompatible FILs for biomedical applications was demonstrated by their great power to solubilize respiratory gases, supporting their use as artificial gas carriers. Additionally, the interaction and the encapsulation of different proteins in FIL aggregates, without comprimising the biological features of the biomolecules, also represents an advance in the application of FILs to pharmaceutical development. Finally, FILs exhibit great ability to be used individually, or in the development of materials to be further applied on the separation and recovery of F-gases, essentially due to their great free volume and gas-FIL enhanced interactions. To conclude, the discussion offered by this chapter highlights the identification of FILs as a novel and endless tool for the design of materials and processes whereas their fluorinated nanosegregated domain in combination with their ionic nature can provide unique features.

Acknowledgements

Authors acknowledge financial support from FCT/MCTES (Portugal), through grant SFRH/BD/130965/2017 and project PTDC/EQU-EQU/29737/2017. This work was also supported by the Associate Laboratory for Green Chemistry - LAQV which is financed by national funds from FCT/MCTES (UIDB/50006/2020).

Author details

Nicole S.M. Vieira, Margarida L. Ferreira, Paulo J. Castro, João M.M. Araújo and Ana B. Pereiro*
LAQV, REQUIMTE, Departamento de Química, Faculdade de Ciências e Tecnologia, Universidade Nova de Lisboa, Caparica, Portugal

*Address all correspondence to: anab@fct.unl.pt

References

[1] Lindstrom AB, Strynar MJ, Libelo, EL. Polyfluorinated compounds: Past, present, and future. Environmental Science & Technology. 2011;45:7954–7961. DOI: 10.1021/es2011622

[2] Emerging chemical risks in europe – PFAS. European Environment Agency. 2019;DOI: 10.2800/486213

[3] Berger R, Resnati G, Metrangolo P, Weber E, Hulliger J. Organic fluorine compounds: a great opportunity for enhanced materials properties. Chemical Society Reviews. 2011;40: 3496–3508. DOI: 10.1039/C0CS00221F

[4] Castro CI, Briceno JC. Perfluorocarbon-based oxygen carriers: review of products and trials. Artificial Organs. 2010;34:622–634. DOI: 10.1111/j.1525-1594.2009.00944.x

[5] Lindstrom AB, Strynar MJ, Libelo, EL, Field JA. Guest comment: Perfluoroalkyl acid focus issue. Environmental Science & Technology. 2011;45:7951–7953. DOI: 10.1021/es202963p

[6] Tsai W.-T, Chen H.-P, Hsien W.-Y. A review of uses, environmental hazards and recovery/recycle technologies of perfluorocarbons (PFCs) emissions from the semiconductor manufacturing processes. Journal of Loss Prevention in the Process Industries. 2002;15:65–75. DOI: 10.1016/S0950-4230(01)00067-5

[7] Melich R, Zorgani A, Padilla F, Charcosset C. Preparation of perfluorocarbon emulsions by premix membrane emulsification for Acoustic Droplet Vaporization (ADV) in biomedical applications. Biomedical Microdevices. 2020;22:62. DOI: 10.1007/s10544-020-00504-5

[8] Choi M, Park S, Park K, Jeong H, Hong J. Nitric oxide delivery using biocompatible perfluorocarbon microemulsion for antibacterial effect. ACS Biomaterials Science & Engineering. 2019;5:1378–1383. DOI: 10.1021/acsbiomaterials.9b00016

[9] Choi H, Choi W, Kim J, Kong WH, Kim KS, Kim C, Hahn, SK. Multifunctional nanodroplets encapsulating naphthalocyanine and perfluorohexane for bimodal image-guided therapy. Biomacromolecules. 2019;20:3767–3777. DOI: 10.1021/acs.biomac.9b00842

[10] Fernandes DA, Kolios MC. Near-infrared absorbing nanoemulsions as nonlinear ultrasound contrast agents for cancer theranostic. Journal of Molecular Liquids. 2019;287:110848. DOI: 10.1016/j.molliq.2019.04.125

[11] Badv M, Alonso-Cantu C, Shakeri A, Hosseinidoust Z, Weitz JI, Didar TF. Biofunctional Lubricant-infused vascular grafts functionalized with silanized bio-inks suppress thrombin generation and promote endothelialization. ACS Biomaterials Science & Engineering. 2019;5:6485–6496. DOI: 10.1021/acsbiomaterials.9b01062

[12] Kohlhauer M, Boissady E, Lidouren F, de Rochefort L, Nadeau M, Rambaud J, Hutin A, Dubuisson R.-M, Guillot G, Pey P, Bruneval P, Fortin-Pellerin E, Sage M, Walti H, Cariou A, Ricard J.-D, Berdeaux A, Mongardon N, Ghaleh B, Micheau P, Tissier R. A new paradigm for lung-conservative total liquid ventilation. EBioMedicine. 2020;50:102365. DOI: 10.1016/j.ebiom.2019.08.026

[13] Courrier HA, Vandamme TF, Krafft MP. Reverse water-in-fluorocarbon emulsions and microemulsions obtained with a fluorinated surfactant. Colloids and Surfaces A Physicochemical and Engineering Aspects. 2004;244:141–148. DOI: 10.1016/j.colsurfa.2004.06.003

[14] Zhu J, Wang Z, Xu X, Xu M, Yang X, Zhang C, Liu J, Zhang F, Shuai X, Wang W, Cao Z. Polydopamine-encapsulated perfluorocarbon for ultrasound contrast imaging and photothermal therapy. Molecular Pharmaceutics. 2020;17:817–826. DOI: 10.1021/acs.molpharmaceut.9b01070

[15] Bakker J, Reihlen A, Meura Lucie, Camboni M, Goldenman G, Lietzmann J. Study for the strategy for a non-toxic environment of the 7th Environment Action Programme. Publications Office of the European Union. 2017;DOI: 10.2779/025

[16] Gyllenhammar I, Berger U, Sundström, M, McCleaf P, Eurén K, Eriksson S, Ahlgren S, Lignell S, Aune M, Kotova N, Glynn A. Influence of contaminated drinking water on perfluoroalkyl acid levels in human serum – A case study from Uppsala, Sweden. Environmental Research. 2015; 140:673–683. DOI: 10.1016/j.envres.2015.05.019

[17] Naile JE, Khim JS, Wang T, Chen C, Luo W, Kwon B.-O, Park J, Koh C.-H, Jones PD, Lu Y, Giesy JP. Perfluorinated compounds in water, sediment, soil and biota from estuarine and coastal areas of Korea. Environmental Pollution. 2010;5: 1237–1244. DOI: 10.1016/j.envpol.2010.01.023

[18] Noorlander CW, van Leeuwen SPJ, te Biesebeek JD, Mengelers MJB, Zeilmaker MJ. Levels of perfluorinated compounds in food and dietary intake of PFOS and PFOA in The Netherlands. Journal of Agricultural and Food Chemistry. 2011;59:7496–7505. DOI: 10.1021/jf104943p

[19] Rogers RD, Seddon KR. Ionic liquids-Solvents of the future?. Science. 2003;302:792–793. DOI: 10.1126/science.1090313

[20] Qiao Y, Ma W, Theyssen N, Chen C, Hou Z. Temperature-responsive ionic liquids: Fundamental behaviors and catalytic applications. Chemical Reviews. 2017;117:6881–6928. DOI: 10.1021/acs.chemrev.6b00652

[21] Watanabe M, Thomas ML, Zhang S, Ueno K, Yasuda T, Dokko K. Application of ionic liquids to energy storage and conversion materials and devices. Chemical Reviews. 2017;117:7190–7239. DOI: 10.1021/acs.chemrev.6b00504

[22] Ventura SPM, A. e Silva F, Quental MV, Mondal D, Freire MG, Coutinho JAP. Ionic-liquid-mediated extraction and separation processes for bioactive compounds: Past, present, and future trends. Chemical Reviews. 2017; 117:6984–7052. DOI: 10.1021/acs.chemrev.6b00550

[23] Egorova KS, Gordeev EG, Ananikov VP. Biological activity of ionic liquids and their application in pharmaceutics and medicine. Chemical Reviews. 2017;117:7132–7189. DOI: 10.1021/acs.chemrev.6b00562

[24] Marrucho IM, Branco LC, Rebelo LPN. Ionic liquids in pharmaceutical applications. Annual Review of Chemical and Biomolecular Engineering. 2014;5:527–546. DOI: 10.1146/annurev-chembioeng-060713-040024

[25] Araújo JMM, Florindo C, Pereiro AB, Vieira NSM, Matia AA, Duarte CMM, Rebelo LPN, Marrucho IM. Cholinium-based ionic liquids with pharmaceutically active anions. RSC Advances. 2014;4:28126–28132. DOI: 10.1039/c3ra47615d

[26] Earle MJ, Esperança JMSS, Gilea MA, Canongia Lopes JN, Rebelo LPN, Magee JW, Seddon KR, Widegren JA. The distillation and volatility of ionic liquids. Nature. 2006; 439:831–834. DOI: 10.1038/nature04451

[27] Plechkova NV, Seddon KR. Applications of ionic liquids in the

chemical industry. Chemical Society Reviews. 2008;37:123–150. DOI: 10.1039/b006677j~

[28] Egorova KS, Ananikov VP. Toxicity of Ionic Liquids: Eco(cyto)activity as complicated, but unavoidable parameter for task-specific optimization. ChemSusChem. 2014;7:336–360. DOI: 10.1002/cssc.201300459

[29] Lei Z, Chen B, Koo Y.-M, MacFarlane DR. Introduction: Ionic liquids. Chemical Reviews. 2017;117: 6633–6635. DOI: 10.1021/acs. chemrev.7b00246

[30] Pereiro AB, Araújo JMM, Martinho S, Alves F, Nunes S, Matias, A, Duarte CMM, Rebelo, Rebelo LPN, Marrucho IM. Fluorinated ionic liquids: Properties and applications. ACS Sustainable Chemistry & Engineering. 2013;1:427–439. DOI: 10.1021/ sc300163n

[31] Pereiro AB, Araújo JMM, Esperança JMSS, Rebelo LPN. Surfactant fluorinated ionic liquids. In: Eftekhari A editor. Ionic Liquid Devices. Smart Materials No. 28: Royal Society of Chemistry; 2018. p. 79–102. DOI: 10.1039/9781788011839

[32] Tindale JJ, Na C, Jennings MC, Ragogna PJ. Synthesis and characterization of fluorinated phosphonium ionic liquids. Canadian Journal of Chemistry. 2007;85:660–667. DOI: 10.1139/V07-035

[33] Pereiro AB, Pastoriza-Gallego MJ, Shimizu K, Marrucho IM, Canongia Lopes JN, Piñeiro MM, Rebelo LPN. On the Formation of a Third, Nanostructured Domain in Ionic Liquids. The Journal of Physical Chemistry B. 2013;117:10826–10833. DOI: 10.1021/jp402300c

[34] Pereiro AB, Araújo JMM, Teixeira FS, Marrucho IM, Piñeiro MM, Rebelo LPN. Aggregation behavior and total miscibility of fluorinated ionic liquids in water. Langmuir. 2015;31: 1283–1295. DOI: 10.1021/la503961h.

[35] Teixeira FS, Vieira NSM, Cortes OA, Araújo JMM, Marrucho IM, Rebelo LPN, Pereiro AB. Phase equilibria and surfactant behavior of fluorinated ionic liquids with water. The Journal of Chemical Thermodynamics. 2015;82:99–107. DOI: 10.1016/j.jct.2014.10.021

[36] Vieira NSM, Stolte S, Araújo JMM, Rebelo LPN, Pereiro AB, Markiewicz M. Acute aquatic toxicity and biodegradability of fluorinated ionic liquids. ACS Sustainable Chemistry & Engineering. 2019;7:3733–3741. DOI: 10.1021/acssuschemeng.8b03653

[37] Vieira NSM, Bastos JC, Rebelo LPN, Matias, A, Araújo JMM, Pereiro AB. Human cytotoxicity and octanol/water partition coefficients of fluorinated ionic liquids. Chemosphere. 2019;216: 576e586. DOI: 10.1016/j. chemosphere.2018.10.159

[38] Prikhod'ko SA, Shabalin AY, Shmakov MM, Bardin VV, Adonin NY. Ionic liquids with fluorine-containing anions as a new class of functional materials: features of the synthesis, physicochemical properties, and use. Russian Chemical Bulletin. 2020;69:17–31. DOI: 1066–5285/20/6901–0017

[39] Tong B, Chen X, Chen L, Zhou Z, Peng Z. Engineering solid electrolyte interphase in lithium metal batteries by employing an ionic liquid ether double-solvent electrolyte with Li[(CF_3SO_2)(n-$C_4F_9SO_2$)N] as the salt. ACS Applied Energy Materials. 2018;1:4426–4431. DOI: 10.1021/acsaem.8b00821.

[40] Lepre LF, Andre D, Denis-Quanquin S, Gautier A, Pádua AAH., Gomes MC. Ionic liquids can enable the recycling of fluorinated greenhouse gases. ACS Sustainable Chemistry & Engineering. 2019;7:16900–16906. DOI: 10.1021/acssuschemeng.9b04214

[41] Rufino-Felipe E, Valdes H, German-Acacio JM, Reyes-Marquez V, Morales-Morales D. Fluorinated N-Heterocyclic carbene complexes. Applications in catalysis. Journal of Organometallic Chemistry. 2020;921:121364. DOI: 10.1016/j.jorganchem.2020.121364

[42] Vieira NSM, Reis PM, Shimizu K, Cortes AO, Marrucho IM, Araújo JMM, Esperança JMSS, Canongia Lopes JN, Pereiro AB, Rebelo LPN. A thermophysical and structural characterization of ionic liquids with alkyl and perfluoroalkyl side chains. RSC Advances. 2015;5:65337–65350. DOI: 10.1039/C5RA13869H

[43] Luís A, Shimizu K, Araújo JMM, Carvalho PJ, Lopes-da-Silva JA, Canongia Lopes JN, Rebelo LPN, Coutinho JAP, Freire MG, Pereiro AB. Influence of nanosegregation on the surface tension of fluorinated ionic liquids. Langmuir. 2016;32:6130–6139. DOI: 10.1021/acs.langmuir.6b00209

[44] Teles ARR, Correia H, Maximo GJ, Rebelo LPN, Freire MG, Pereiro AB, Coutinho JAP. Solid–liquid equilibria of binary mixtures of fluorinated ionic liquids. Physical Chemistry Chemical Physics. 2016;18:25741–25750. DOI: 10.1039/C6CP05372F

[45] Vieira NSM, Luís A, Reis PM, Carvalho PJ, Lopes-da-Silva JA, Esperança JMSS, Araújo JMM, Rebelo LPN, Freire MG, Pereiro AB. Fluorination effects on the thermodynamic, thermophysical and surface properties of ionic liquids. The Journal of Chemical Thermodynamics. 2016;97:354–361. DOI: 10.1016/j.jct.2016.02.013

[46] Ferreira ML, Pastoriza-Gallego MJ, Araújo JMM, Canongia Lopes JN, Rebelo LPN, Piñeiro MM, Shimizu K, Pereiro AB. Influence of nanosegregation on the phase behavior of fluorinated ionic liquids. The Journal of Physical Chemistry C. 2017;121:5415–5427. DOI: 10.1021/acs.jpcc.7b00516

[47] Pereiro AB, Llovell F, Araújo JMM, Santos ASS, Rebelo LPN, Piñeiro MM, Vega LF. Thermophysical Characterization of ionic liquids based on the perfluorobutanesulfonate anion: experimental and soft-SAFT modeling results. ChemPhysChem. 2017;18:1–13. DOI: 10.1002/cphc.201700327

[48] Ferreira ML, Araújo JMM, Pereiro AB, Vega LF. Insights into the influence of the molecular structures of fluorinated ionic liquids on their thermophysical properties. A soft-SAFT based approach. Physical Chemistry Chemical Physics. 2019;21:6362–6380. DOI: 10.1039/C8CP07522K

[49] Ferreira ML, Llovell F, Vega LF, Pereiro AB, Araújo JMM. Systematic study of the influence of the molecular structure of fluorinated ionic liquids on the solubilization of atmospheric gases using a soft-SAFT based approach. Journal of Molecular Liquids. 2019;294:111645. DOI: 10.1016/j.molliq.2019.111645

[50] Ferreira ML, Araújo JMM, Vega LF, Llovell F, Pereiro AB. Functionalization of fluorinated ionic liquids: A combined experimental-theoretical study. Journal of Molecular Liquids. 2020;302:112489. DOI: 10.1016/j.molliq.2020.112489

[51] Bastos JC, Carvalho SF, Welton T, Canongia Lopes JN, Rebelo LPN, Shimizu K, Araújo JMM, Pereiro AB. Design of task-specific fluorinated ionic liquids: nanosegregation versus hydrogen-bonding ability in aqueous solutions. Chemical Communications. 2018;54:3524–3527. DOI: 10.1039/C8CC00361K

[52] Vieira NSM, Bastos JC, Hermida-Merino C, Pastoriza-Gallego MJ, Rebelo LPN, Piñeiro MM, Araújo JMM, Pereiro AB. Aggregation and phase equilibria of fluorinated ionic liquids.

Journal of Molecular Liquids. 2019;285: 386–396. DOI: 10.1016/j.molliq.2019.04.086

[53] Gupta S, Olson JD. Industrial needs in physical properties. Industrial & Engineering Chemistry Research. 2003;42:6359–6374. DOI: 10.1021/ie030170v

[54] Ueno K, Tokuda H, Watanabe M. Ionicity in ionic liquids: correlation with ionic structure and physicochemical properties. Physical Chemistry Chemical Physics. 2010;12:1649–1658. DOI: 10.1039/B921462N

[55] Tariq M, Forte PAS, Costa Gomes MF, Canongia Lopes JN, Rebelo LPN. Densities and refractive indices of imidazolium- and phosphonium-based ionic liquids: Effect of temperature, alkyl chain length, and anion. The Journal of Chemical Thermodynamics. 2009;41:790–798. DOI: 10.1016/j.jct.2009.01.012

[56] Santos CS, Baldelli S, Gas-liquid interface of room-temperature ionic liquids. Chemical Society Reviews. 2010;39:2136–2145. DOI: 10.1039/b921580h.

[57] Szajdzinska-Pietek E, Wolszczak M. Time-resolved fluorescence quenching study of aqueous solutions of perfluorinated surfactants with the use of protiated luminophore and quencher. Langmuir. 2000;16:1675–1680. DOI: 10.1021/LA990981X

[58] González-Pérez A, Ruso JM, Prieto G, Sarmiento F. Apparent molar quantities of sodium octanoate in aqueous solutions. Colloid and Polymer Science. 2004;282:1133–1139. DOI: 10.1007/s00396-003-1047-2

[59] López-Fontán JL, Sarmiento F, Schulz PC. The aggregation of sodium perfluorooctanoate in water. Colloid and Polymer Science. 2004;283:862–871. DOI: 10.1007/s00396-004-1228-7

[60] Ventura SPM, Gonçalves AMM, Sintra T, Pereira JL, Gonçalves F, Coutinho JAP. Designing ionic liquids: the chemical structure role in the toxicity. Ecotoxicology. 2013;22:1–12. DOI: 10.1007/s10646-012-0997-x

[61] Petkovic M, Ferguson JL, Gunaratne HQN, Ferreira R, Leitão MC, Seddon KR, Rebelo LPN, Pereira CS. Novel biocompatible cholinium-based ionic liquids-toxicity and biodegradability. Green Chemistry. 2010;12:643–649. DOI: 10.1039/B922247B

[62] Patinha DJS, Tomé LC, Florindo C, Soares HR, Coroadinha AS, Marrucho IM. New low-toxicity cholinium-based ionic liquids with perfluoroalkanoate anions for aqueous biphasic system implementation. ACS Sustainable Chemistry & Engineering. 2016;4:2670–2679. DOI: 10.1021/acssuschemeng.6b00171

[63] Gal N, Malferarri D, Kolusheva S, Galletti P, Tagliavini E, Jelinek R. Membrane interactions of ionic liquids: possible determinants for biological activity and toxicity. Biochimica et Biophysica Acta (BBA) – Biomembranes. 2012;1818:2967–2974. DOI: 10.1016/j.bbamem.2012.07.025

[64] Kumar RA, Papaïconomou N, Lee J, Salminen J, Clark DS, Prausnitz JM. In vitro cytotoxicities of ionic liquids: effect of cation rings, functional groups, and anions. Environmental Toxicology. 2008;24:388–395. DOI: 10.1002/tox

[65] Lungwitz R, Strehmel V, Spange S. The dipolarity/polarisability of 1-alkyl-3-methylimidazolium ionic liquids as function of anion structure and the alkyl chain length. New Journal of Chemistry. 2010;34:1135–1140. DOI: 10.1039/B9NJ00751B

[66] Kim, M., Li, L.Y., Grace, J.R., Yue, C., Selecting reliable physicochemical properties of perfluoroalkyl and

polyfluoroalkyl substances (PFASs) based on molecular descriptors. Environmental Pollution. 2015;196:462–472. DOI: 10.1016/j.envpol.2014.11.008

[67] Docherty KM, Dixon JK., Kulpa CF. Biodegradability of imidazolium and pyridinium ionic liquids by an activated sludge microbial community. Biodegradation. 2007;18:481–493. DOI: 10.1007/s10532-006-9081-7

[68] Zhang C, Wang H, Malhotra SV, Dodge CJ, Francis AJ. Biodegradation of pyridinium-based ionic liquids by an axenic culture of soil Corynebacteria. Green Chemistry. 2010;12:851–858. DOI: 10.1039/B924264C

[69] Busch MP, Kleinman SH, Nemo GJ. Current and emerging infectious risks of blood transfusions. The Journal of the American Medical Association. 2003;289: 959–962. DOI: 10.1001/jama.289.8.959

[70] Martinho S, Araújo JMM, Rebelo LPN, Pereiro AB, Marrucho IM. The Journal of Chemical Thermodynamics. 2013;64:71. DOI: 10.1016/j.jct.2013.04.019

[71] Pereiro AB, Tomé LC, Martinho S, Rebelo LPN, Marrucho IM. Industrial & Engineering Chemistry Research. 2013; 52:4994–5001. DOI: 10.1021/ie4002469

[72] Fujita K, MacFarlane DR, Forsyth M, Yoshizawa-Fujita M, Murata K, Nakamura N, Ohno H. Solubility and stability of cytochrome c in hydrated ionic Liquids: effect of oxo acid residues and kosmotropicity. Biomacromolecules.2007;8:2080–2086. DOI: 10.1021/bm070041o

[73] Moniruzzaman M, Tamura M, Tahara Y, Kamiya N, Goto M. Ionic liquid- in-oil microemulsion as a potential carrier of sparingly soluble drug: characterization and cytotoxicity evaluation. International Journal of Pharmaceutics. 2010;400:243–250. DOI: 10.1016/j.ijpharm.2010.08.034

[74] Alves M, Vieira NSM, Rebelo LPN, Araújo JMM, Pereiro AB, Archer, M. Fluorinated ionic liquids for protein drug delivery systems: Investigating their impact on the structure and function of lysozyme. International Journal of Pharmaceutics. 2017;526: 309–320. DOI: 10.1016/j.ijpharm.2017.05.002

[75] Alves M, Araújo JMM, Martins IC, Pereiro AB, Archer, M. Insights into the interaction of bovine serum albumin with surface-active ionic liquids in aqueous solution. Journal of Molecular Liquids. 2020;114537. DOI: 10.1016/j.molliq.2020.114537

[76] Vieira NSM, Castro PJ, Marques DF, Araújo JMM, Pereiro AB. Tailor-made fluorinated ionic liquids for protein delivery. Nanomaterials. 2020;10:1594. Doi:10.3390/nano10081594

[77] Ibraheem D, Elaissari A, Fessi H. Administration strategies for proteins and peptides. International Journal of Pharmaceutics. 2014;477:578–589. DOI: 10.1016/j.ijpharm.2014.10.059

[78] Dai C, Wanga B, Zhao H. Microencapsulation peptide and protein drugs delivery system. Colloids and Surfaces B: Biointerfaces. 2005;41: 117–120. DOI: 10.1016/j.colsurfb.2004.10.032

[79] Mishra H, Chauhan V, Kumar K, Teotia D. A comprehensive review on liposomes: A novel drug delivery system. Journal of Drug Delivery and Therapeutics. 2018;8:400–404. DOI: 10.22270

[80] Abeyrathne EDNS, Lee HY, Ahn DU. Egg white proteins and their potential use in food processing or as nutraceutical and pharmaceutical agents —A review. Poultry Science. 2013;92: 3292–3299. DOI: 10.3382/ps.2013-03391

[81] Bujacz A. Structures of bovine, equine and leporine serum albumin.

Acta Crystallographica Section D. 2012; 68:1278–1289. DOI: 10.1107/S0907444912027047

[82] F. Geng, L. Zheng, J. Liu, L. Yu, C. Tung, Interactions between a surface active imidazolium ionic liquid and BSA. Colloid and Polymer Science. 2009;287:1253–1259. DOI: 10.1007/s00396-009-2085-1

[83] Sosa JE, Ribeiro RPPL, Castro PJ, Mota JPB, Araújo JMM, Pereiro AB. Absorption of fluorinated greenhouse gases using fluorinated ionic liquids. Industrial & Engineering Chemistry Research. 2019;58:20769–20778. DOI: 10.1021/acs.iecr.9b04648

[84] Shiflett MB, Yokozeki A. Solubility and diffusivity of hydrofluorocarbons in room-temperature ionic liquids. AIChE Journal. 2006;52:1205–1219. DOI: 10.1002/aic.10685

[85] Shiflett MB, Yokozeki A. Binary vapor-liquid and vapor- liquid-liquid equilibria of hydrofluorocarbons (HFC-125 and HFC-143a) and hydrofluoroethers (HFE-125 and HFE-143a) with ionic liquid [Emim][Tf2N]. Journal of Chemical & Engineering Data. 2008;53:492–497. DOI: 10.1021/je700588d

[86] Shiflett, M. B.; Yokozeki, A. vapor-liquid-liquid equilibria of pentafluoroethane and ionic liquid [Bmim][PF6] mixtures studied with the volumetric method. The Journal of Physical Chemistry B. 2006;110:14436–14443. DOI: 10.1021/jp062437k

[87] Shiflett MB, Harmer MA, Junk CP, Yokozeki A. solubility and diffusivity of difluoromethane in room-temperature ionic liquids. Journal of Chemical & Engineering Data. 2006;51:483–495. DOI: 10.1021/je050386z

[88] Ren W, Scurto AM. Phase equilibria of imidazolium ionic liquids and the refrigerant gas, 1,1,1,2-Tetrafluoroethane (R-134a). Fluid Phase Equilibria. 2009;286:1–7. DOI: 10.1016/j.fluid.2009.07.007

[89] Lepre LF, Pison L, Otero I, Gautier A, Dévemy J, Husson P, Pádua AAH, Gomes MC. Using hydrogenated and perfluorinated gases to probe the interactions and structure of fluorinated ionic liquids. Physical Chemistry Chemical Physics. 2019;21: 8865–8873. DOI:10.1039/c9cp00593e

[90] Lepre LF, Andre D, Denis-Quanquin S, Gautier A, Pádua AAH, Gomes MC. Ionic liquids can enable the recycling of fluorinated greenhouse gases. ACS Sustainable Chemistry & Engineering. 2019;7:16900–16906. DOI: 10.1021/acssuschemeng.9b04214

[91] Jovell D, Gómez SB, Zakrzewska ME, Nunes AVM, Araújo JMM. Pereiro AB, Llovell F. Insight on the solubility of R134a in fluorinated ionic liquids and deep eutectic solvents. Journal of Chemical & Engineering Data. 2020;65:4956–4969. DOI: 10.1021/acs.jced.0c00588

[92] Liu X, He M, Lv N, Qi X, Su C. Vapor-liquid equilibrium of three hydrofluorocarbons with [HMIM][Tf2N]. Journal of Chemical & Engineering Data. 2015;60:1354–1361. DOI: 10.1021/je501069b

[93] Dong L, Zheng D, Sun G, Wu X. Vapor—liquid equilibrium measurements of difluoromethane + [Emim]OTf,difluoromethane + [Bmim]OTf, difluoroethane + [Emim]OTf, and difluoroethane + [Bmim]OTf systems. Journal of Chemical & Engineering Data. 2011;56:3663–3668. DOI: 10.1021/je2005566

[94] Sosa JE, Santiago R, Hospital-Benito D, Gomes MC, Araújo JMM, Pereiro AB, Palomar J. Process evaluation of fluorinated ionic liquids as F-gas absorbents. Environmental Science & Technology. 2020;54:12784–12794. DOI: 10.1021/acs.est.0c05305

[95] Leron RB, Li M. Solubility of carbon dioxide in a choline chloride-ethylene glycol based deep eutectic solvent. Thermochimica Acta. 2013;551:14–19. DOI: 10.1016/j.tca.2012.09.041

[96] Leron RB, Caparanga A, Li M. Carbon dioxide solubility in a deep eutectic solvent based on choline chloride and urea at T = 303.15–343.15K and moderate pressures. Journal of the Taiwan Institute of Chemical Engineers. 2013;44:879–885. DOI: 10.1016/j.jtice.2013.02.005

[97] Francisco M, van den Bruinhorst A, Zubeir LF, Peters CJ, Kroon MCA. New low transition temperature mixture (LTTM) formed by choline chloride +lactic acid: Characterization as solvent for CO2 capture. Fluid Phase Equilibria. 2013;340:77–84. DOI: 10.1016/j.fluid.2012.12.001

[98] Yang D, Hou M, Ning H, Zhang J, Ma J, Yang G, Han B. Efficient SO2 Absorption by renewable choline chloride–glycerol deep eutectic solvents. Green Chemistry. 2013;15:2261–2265. DOI: 10.1039/C3GC40815A

[99] Castro PJ, Redondo AE, Sosa JE, Zakrzewska ME, Nunes AVM, Araújo JMM, Pereiro AB. Absorption of fluorinated greenhouse gases in deep eutectic solvents. Industrial & Engineering Chemistry Research. 2020;59:13246–13259. DOI: 10.1021/acs.iecr.0c01893

Chapter 3

Group of Uniform Materials Based on Organic Salts (GUMBOS): A Review of Their Solid State Properties and Applications

Rocío L. Pérez, Caitlan E. Ayala and Isiah M. Warner

Abstract

Ionic liquids (ILs) are defined as organic salts with melting points below 100 °C. Such ionic compounds are typically formed using bulky cations and/or bulky anions in order to produce liquids or lower melting solids. ILs have been widely explored in several research areas including catalysis, remediation, solvents, separations, and many others. The utility of such compounds has also been recently broadened to include solid phase ionic materials. Thus, researchers have pushed the boundaries of ILs chemistry toward the solid state and have hypothesized that valuable properties of ILs can be preserved and fine-tuned to achieve comparable properties in the solid state. In addition, as with ILs, tunability of these solid-phase materials can be achieved through simple counterion metathesis reactions. These solid-state forms of ILs have been designated as a *g*roup of *u*niform *m*aterials *b*ased on *o*rganic *s*alts (GUMBOS). In contrast to ILs, these materials have an expanded melting point range of 25 to 250 °C. In this chapter, we focus on recent developments and studies from the literature that provide for fine tuning and enhancing properties through transformation and recycling of diverse ionic compounds such as dyes, antibiotics, and others into solid state ionic materials of greater utility.

Keywords: GUMBOS, solid-phase materials, nanoGUMBOS, fluorescence, sensors

1. Introduction

In recent years, many different kinds of materials and techniques have been developed for improved analytical measurements [1–7]. However, in order to be generally applicable, most materials should have several key properties. These desired properties include, but are not limited to (1) simplicity of preparation (e.g. development involves simply mixing two chemical solutions), (2) tunability (easy introduction of uniform multifunctionality through simple variations), and (3) limited or no toxicity (can be easily designed using materials that are already Food and Drug Administration (FDA) approved). As an example of the latter material, the near infrared (NIR) dye, indocyanine green, for near infrared fluorescence measurements has received early approval by the FDA [7].

On the basis of the above considerations, a wide variety of materials and nanomaterials have been developed and employed for bioanalytical and environmental

IntechOpen

measurements. In regard to nanomaterials, studies reveal that in general primary properties such as spectra, colorimetric response, and magnetism are size dependent and somewhat tunable. Some of these materials, including carbon dots and silicon dots, exhibit very low cytotoxicities. However, other nanomaterials such as carbon nanotubes and quantum dots have considerably higher toxicities. In some cases, e.g. P-dots and nanogels [8–10], toxicity depends on the type of polymer used. Aqueous co-ordination complexes are another category of materials and nanomaterials with variable toxicities that have recently been used for analytical and environmental applications [11, 12].

We believe that when one does an exhaustive examination of the literature and considers the inherent properties identified above for improved analytical measurements, a logical conclusion is that ILs, GUMBOS, and nanomaterials derived from GUMBOS (nanoGUMBOS) represent novel classes of materials that best satisfies all of the above properties. Both ILs and GUMBOS are based on use of organic salts. Examples of typical ions used in these salts (ILs and GUMBOS) are shown in **Figure 1**.

These materials are continually being explored for improved analytical measurements. In fact, the literature on development of novel methodologies based on use of ionic liquids (ILs), a group of uniform materials based on organic salts (GUMBOS), and nanoGUMBOS is increasing at an ever-expanding rate. For example, numerous studies from the literature can be cited for utility of such materials in diverse areas such as antibiotics [13–15], cancer therapy [16–24], hydrogels [25, 26], cellular imaging [27, 28], chirality [29–33], dye-sensitized solar cells (DSSCs) [34–36], extractions [37–39], gel electrophoresis [40, 41], detection of reactive oxygen species [42], liquid crystals [43], mass spectrometry [44], nanomaterials [45–52], optoelectronics [53–56], sensors [57–61], separation science [62, 63], spectroscopy [64–67], volatile organic compounds (VOCs) [68–75], as well as a number of patents and patent applications [76–78]. **Figure 2** provides an abbreviated summary of numerous applications of ILs and GUMBOS.

We note that GUMBOS and nanoGUMBOS are solid phase organic salts (m.p. > 25 °C and < 250 °C) and ILs are typically liquids or low melting solids (m.p. < 100 °C). See **Figure 3** below for differentiation between ILs and GUMBOS in terms of melting points. Therefore, some GUMBOS (and nanoGUMBOS) materials fit into the general category of frozen ILs since ILs from 100 °C down to 25 °C are solids. However, many GUMBOS materials are outside the generally accepted temperature range for ILs. Accordingly, a new, more general term of GUMBOS, as defined above, was adopted to apply to this entire class of solid phase organic salts.

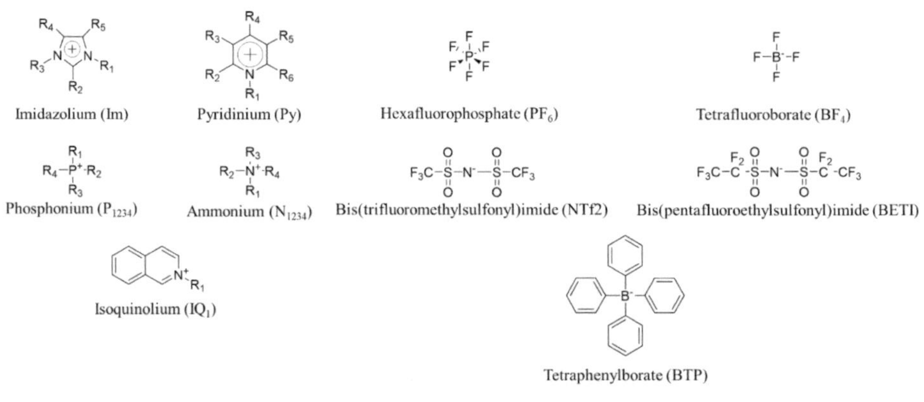

Figure 1.
Typical cations and anions used for ILs/GUMBOS production.

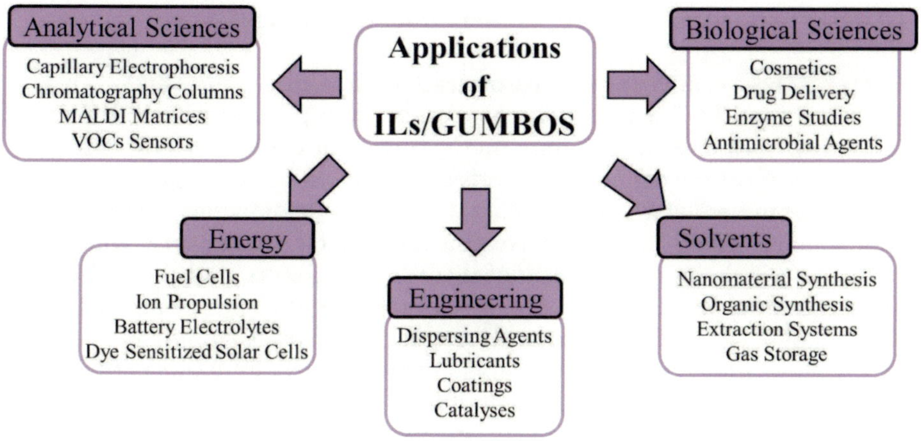

Figure 2.
Applications of ILs/GUMBOS in different research areas.

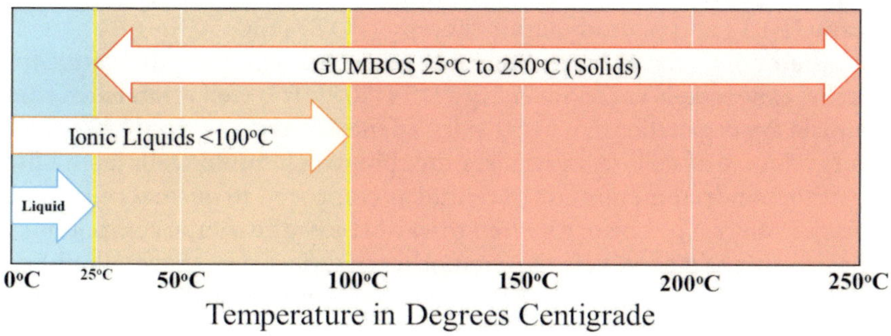

Figure 3.
Melting points range of ILs and GUMBOS.

To date, numerous strategies for this kind of chemistry have been developed. In this chapter, we desire to discuss some of these applications in detail, particularly as applied to the general area of analytical and environmental chemistry.

2. Biological applications

ILs have been recognized for their properties such as non-volatility, viscosity, negligible vapor pressure, high ionic strength, thermal stability, and low toxicity, among others [79]. As a result of these important properties, ILs were initially designated as green and designer solvents (i.e. first generation ILs) [80]. Eventually, due to their high tunability, new ILs were strategically designed for a variety of functional materials, including lubricants, catalysts, energy materials, etc. [81–86]. These types of ILs are known as second generation ILs. Finally, major interest has focused on development of new ILs (third generation ILs) for biological applications to achieve biocompatible and low toxic compounds through use of bio-counterions [87, 88]. Moving forward, more attention from the scientific community has focused on development, or recycling of, various molecules into solid phase materials (frozen ILs or GUMBOS) for several biological applications [17, 89–91]. In this section, the use of frozen ILs and GUMBOS for biological applications such as cancer and antibiotic therapies are discussed.

2.1 GUMBOS and nanoGUMBOS as chemotherapeutic agents

Cancer is the second leading cause of death in the United States and is a major health concern worldwide [92]. Treatment of cancer typically includes surgery, radiotherapy, hormone therapy, immunotherapy and/or chemotherapy [93]. Effectiveness of these treatments depends upon several factors, such as stage of cancer at the moment of diagnosis, general health of the individual, size and type of tumor, among others. In general, treatment of a person with cancer will involve a combination of therapies as a result of these several factors, and chemotherapy is the most commonly employed treatment. Unfortunately, chemotherapy will often be accompanied with several adverse effects, such as nausea, vomiting, diarrhea, fatigue, malnutrition, anemia, hepatotoxicity, nephrotoxicity, among others [94–96]. These side effects are the result of high toxicity of typical chemotherapeutic agents that generally lack selectivity toward carcinogenic cells. For all of these reasons, over the past decades major attention has focused on developing new chemotherapeutic agents that are selectively toxic to cancer cells [97–99]. Moreover, investigations have also focused on early detection methods that involve use of tumor-targeting dyes, as well as near infrared (NIR) dyes for detection, photothermal therapy (PTT) and photodynamic therapy (PDT) [100–103].

Due to their relatively high division rate and subsequent growth relative to normal cells, cancer cells use more energy [104–106]. It is well established that the mitochondria are organelles that synthesize adenosine triphosphate (ATP), which is the energy source of cells. As a result, mitochondria in tumor cells have a higher negative mitochondrial membrane potential as compared to normal cells. For this reason, major interest has been directed toward study of cationic compounds as well as positively charged vesicles as chemotherapeutic agents. Several publications from the literature document that these type of compounds are attracted to, and accumulate more selectively, in this organelle of cancer cells, resulting in disruption of ATP synthesis and subsequent induction of cell death [16, 17, 22, 89, 107, 108].

Cationic rhodamine dyes have been studied as mitochondrial targeting agents as early as the 1970s [109–114]. Furthermore, studies with rhodamine dyes demonstrate that these dyes are toxic to cells above certain concentrations [115, 116]. In contrast, it has been previously reported that hydrophobicity of drugs may improve cellular uptake and distribution inside cancer cells [117]. For this reason, Magut et al. hypothesized that counterion variation in rhodamine 6G dye ($[R6G]^+$) may tune its hydrophobicity [22]. In this regard, four anions: ascorbate ($[Asc]^-$), trifluoremethanesulfonate ($[OTf]^-$), tetraphenylborate ($[TPB]^-$) and bis(perfluoroethylsulfonyl)imide ($[BETI]^-$) were employed to synthesize, through a simple metathesis reaction, four R6G-based GUMBOS. Relative hydrophobicities for each GUMBOS were determined, and the following trend in increasing hydrophobicity from [R6G][Asc] < [R6G][OTf] < [R6G][TPB] < [R6G][BETI] was observed in this study. Clearly, anion variation affected and tuned hydrophobicity, along with other physico-chemical properties, of the parent dye. The low water solubility of these compounds allowed synthesis of nanoGUMBOS through a simple reprecipitation method. *In vitro* cellular cytotoxicity of these GUMBOS and nanoGUMBOS towards normal breast cells (Hs578T), hormone-independent human breast adenocarcinoma (MDA-MB-231) and hormone-dependent human breast adenocarcinoma (MCF7) cell lines using an MTT assay was evaluated. Interestingly, evaluation of results obtained for [R6G][Asc] and [R6G][OTf] showed that these GUMBOS were highly toxicity toward both normal and cancer cell lines. Similar behavior was observed with the parent compound, [R6G][Cl]. This trend was explained through similar water solubilities of these compounds. In contrast, cytotoxicity results obtained for [R6G][BETI] and [R6G][TPB] indicated that these compounds were more selectively toxic toward cancer cell lines than normal cells. Moreover, these GUMBOS were more toxic against MDA-MB-231 cancer cells, which were the most aggressive cancer cell lines evaluated using 50% inhibition concentration (IC_{50}) values of 11.4 and 12.2 μM for [R6G][BETI] and [R6G][TPB], respectively. Additionally, confocal microscopy studies demonstrated that these nanoGUMBOS were localized inside the mitochondria of cancer cells, which resulted in decreased synthesis of ATP.

Following this study, other researchers focused on evaluating the mechanism of action and internalization of [R6G][BETI] nanoGUMBOS in cells [18]. In that study, Bhattarai and coworkers performed a series of *in vitro* experiments at different incubation temperatures, in the presence of several endocytic inhibitors, as well as in depletion media, to study internalization of [R6G][BETI] nanoGUMBOS. These experiments allowed the investigators to conclude that these nanoGUMBOS were internalized into cancer cells through a clathrin mediated endocytosis pathway. In contrast, these nanoGUMBOS were found to be internalized in normal cells through an independent endocytic route. Interestingly, it was further demonstrated that [R6G][BETI] nanoGUMBOS passed through lysosome vesicles before reaching the mitochondria. For this reason, Bhattarai and coworkers investigated the integrity of these nanoparticles at lysosomic pH (pH = 4) and normal pH (7.4) values using transmission electron microscopy (TEM) and dynamic light scattering (DLS) experiments. Evaluation of these results demonstrated that nanoparticles lost integrity under acidic pH similar to those in the lysosome, thus releasing $[R6G]^+$ that subsequently entered the mitochondria, and inhibited ATP synthesis and eventually causing apoptosis of cancer cells. Internalization in normal cells did not involve entry through the lysosome. Thus, these investigators concluded that this differential internalization route was the primary reason for selectivity of [R6G][BETI] nanoGUMBOS towards cancer cells. Finally, Bhattarai and co-workers evaluated *in vivo* efficacy of nanoGUMBOS in tumor size reduction in an athymic nude mouse model. Evaluation of these results demonstrated that nanoGUMBOS inhibited tumor growth and decreased tumors size by 50% making this material a good candidate for *in vivo* chemotherapeutic applications [18].

It has been previously reported in the literature that there is a strong correlation between size, hydrophobicity, and surface charge of nanomaterials as related to resultant toxicity against cancer cells [118, 119]. Moreover, *in vivo* studies have demonstrated that nanoparticle size has an important effect in increasing cellular uptake into tumor tissue via leaky tumor vasculature through a phenomenon known as enhanced permeability and retention (EPR) effect [120]. In this regard, Hamdan et al. have shown that use of cyclodextrins (CDs) in nanoGUMBOS synthesis results in more uniform and smaller nanoparticles [45]. CDs are cyclic oligosaccharides having conical shapes, with an inner hydrophobic cavity and an external hydrophilic surface. This characteristic structure of CDs allows interaction with some hydrophobic compounds to provide encapsulation and increased water solubility [121, 122]. For this reason, Bhattarai et al. investigated the synthetic procedure of [R6G][BETI] and [R6G][TPB] nanoGUMBOS in the presence of three different CDs: 2-hydroxypropyl-αCD (HP-αCD), 2-hydropropyl-βCD (HP-βCD), and γ-CD in order to optimize nanoparticle size, uniformity and stability [17]. In this report, nanoparticle synthesis was performed by directly mixing stoichiometric quantities of each parent compound in the presence of predetermined concentrations of each CD until synthetic conditions were optimized. These researchers noticed, from TEM and zeta potential data, that CD-templated nanoparticles presented lower size and higher zeta potentials as compared to control nanoGUMBOS (**Figure 4**). In **Table 1**, size, zeta potential and cytotoxicity of CD-templated and control nanoGUMBOS are summarized. Based on evaluation of results obtained

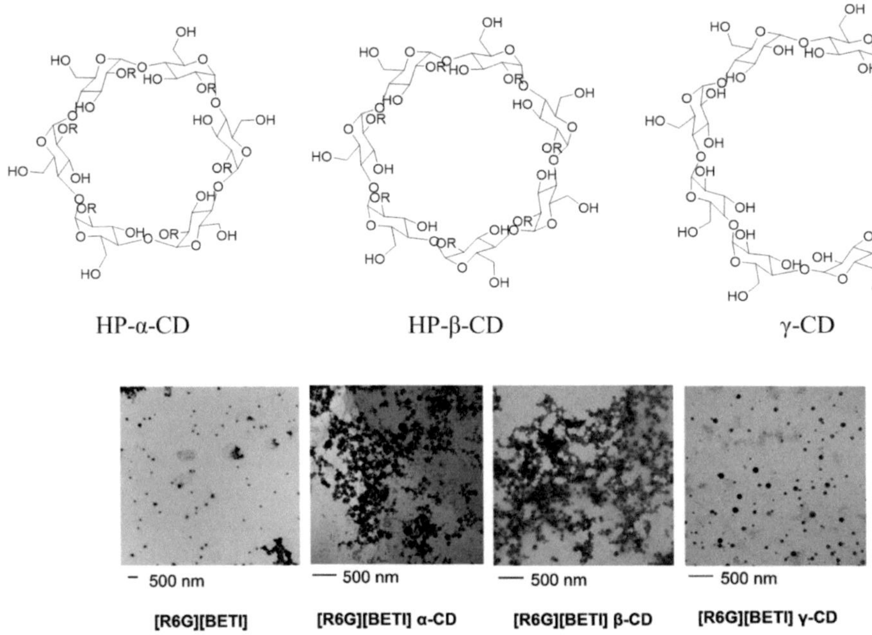

Figure 4.
CDs structures and TEM images of [R6G][BETI] nanoGUMBOS in absence and presence of CDs.

Compound	Size (nm)	Zeta potential (mV)	IC_{50} MDA-MB-231 (μg mL^{-1})	IC_{50} MiaPaca (μg mL^{-1})
[R6G][TPB] control	105 ± 16	−23.1 ± 1.2	7.3 ± 1.1	0.75 ± 0.05
[R6G][TPB] HP-α-CD	55 ± 6	−27.2 ± 1.5	2.6 ± 0.2	0.37 ± 0.03
[R6G][TPB] HP-β-CD	44 ± 4	−29.5 ± 1.1	2.7 ± 0.3	0.39 ± 0.06
[R6G][TPB] γ-CD	69 ± 6	−28.3 ± 0.9	1.4 ± 0.3	0.24 ± 0.04
[R6G][BETI] control	99 ± 12	−24.3 ± 1.2	4.2 ± 0.4	0.45 ± 0.05
[R6G][BETI] HP-α-CD	68 ± 8	−29.0 ± 1.1	1.6 ± 0.3	0.24 ± 0.03
[R6G][BETI] HP-β-CD	66 ± 4	−30.1 ± 0.8	1.7 ± 0.2	0.26 ± 0.04
[R6G][BETI] γ-CD	80 ± 5	−29.8 ± 1.6	2.3 ± 0.4	0.30 ± 0.03

Table 1.
Size, zeta potential, IC_{50} for MDA-MB-231 and MiaPaca cell lines for parent and CD-templated nanoGUMBOS [17].

using cytotoxicity studies performed with MDA-MB-231 and pancreatic cancer (MiaPaca) cell lines, these researchers noticed a significant decrease in IC_{50} values, suggesting that CD-templating enhances toxicity of these nanoparticles.

Another approach from this laboratory involved use of an IR-780 dye to synthesize GUMBOS and nanoGUMBOS [16]. IR-780 is a NIR fluorescent dye that has been studied as a possible theranostic agent since it can be employed as an imaging agent as well as a photothermal and photodynamic agent [123–125]. In this regard, Chen et al. synthesized three IR-780-based GUMBOS through a simple metathesis reaction [16]. Anions evaluated in that study were [Asc]$^-$, [OTf]$^-$, and [BETI]$^-$. Relative hydrophobicity and spectroscopic properties of these GUMBOS were evaluated and compared to the parent compound [IR-780][I]. These researchers found the following hydrophobicity trend: [IR-780][BETI] > [IR-780][I] > [IR-780][OTf] > [IR-780][Asc]. As a result of these larger hydrophobicity values, nanoGUMBOS synthesis was performed through a simple reprecipitation method. Cytotoxicity of [IR-780][BETI] nanoGUMBOS and [IR-780][I] nanomaterials were studied *in vitro* in three different cancer cell lines: MDA-MB-231, MCF7 and MiaPaca using an MTT assay, and IC_{50} values were calculated. Interestingly, IC_{50} for [IR-780][BETI] were lower than IC_{50} values for nanoparticles of the parent compound for all cell lines evaluated. Furthermore, this [IR-780][BETI] nanoGUMBOS presented the lowest IC_{50} values against MDA-MB-231, which was the most invasive and aggressive cancer cell line evaluated. These findings indicate that a simple anion variation in a parent compound can selectively change its cytotoxicity towards cancer cell lines.

Relative cell viability was evaluated for each nanoGUMBOS in normal breast cells. These researchers found that all nanoGUMBOS studied were more selectively cytotoxic against cancer cell lines. Results observed after cellular uptake and fluorescence microscopy studies of each nanomaterial allowed these researchers to conclude that nanoGUMBOS, especially [IR-780][BETI], were internalized and accumulated within the mitochondria in higher amounts than with the parent compound. It has been previously reported in the literature that mitochondrial accumulation of [IR-780][I] is followed by cellular apoptosis [126]. In addition, these researchers investigated nanoGUMBOS as inducers of necrosis or mitochondrial disruptors, by employing a mitochondrial toxicity assay. Evaluation of these results showed that nanoGUMBOS presented behavior similar to [IR-780][I] nanomaterials and acted as mitochondrial toxins by inhibiting oxidative phosphorylation. In summary, nanoGUMBOS synthesized in this work represented great potential as possible chemotherapeutic agents along with a strategic advantage as compared to other reported nanomaterials that require complicated synthetic procedures and labels to increase selectivity against cancer cells [127–130].

In another work, Chen and coworkers evaluated *in vitro* and *in vivo* cytotoxicity and photothermal properties of CD-[IR-780][TPB] complexed nanoGUMBOS [89]. In this work, [IR-780][TPB] GUMBOS were synthesized and nanoGUMBOS were obtained in using HP-β-CD. In this case, CD-[IR-780][TPB] nanoGUMBOS represented larger diameters than nanoGUMBOS without CD (**Figure 5a** and **b**). These results were different from those obtained by Bhattarai et al. [17], in which CDs acted as a template. Based on analyses of TEM and differential scanning calorimetry results, along with computational modelling, the authors demonstrated that a stable complex was formed between HP-β-CD and [IR-780][TPB]. Afterwards, these researchers performed relative cell viability studies in breast cancer cell lines (MDa-MB-231, MCF-7, and Hs578T) and normal breast cell lines (Hs578Bst and HMEC). Based on these results, Chen and partners demonstrated that CD-[IR-780][TPB] nanoGUMBOS were more selective against cancer cells in comparison to [IR-780][I] nanoparticles and [IR-780][TPB] nanoGUMBOS. Moreover, CD-[IR-780][TPB] nanoGUMBOS were not toxic to breast normal cells in the concentration range evaluated.

Additionally, cell viability of CD-[IR-780][TPB] nanoGUMBOS were evaluated using MDA-MB-231 and Hs578T cell lines with NIR laser irradiation (808 nm). This resulted in a further decrease in IC_{50} values for these nanomaterials. Furthermore, these results demonstrate that CD-[IR-780][TPB] nanoGUMBOS represent highly potent chemo- and photothermal therapeutic agents. Finally, Chen et al. also evaluated *in vivo* efficacy and PTT activity of nanoGUMBOS and

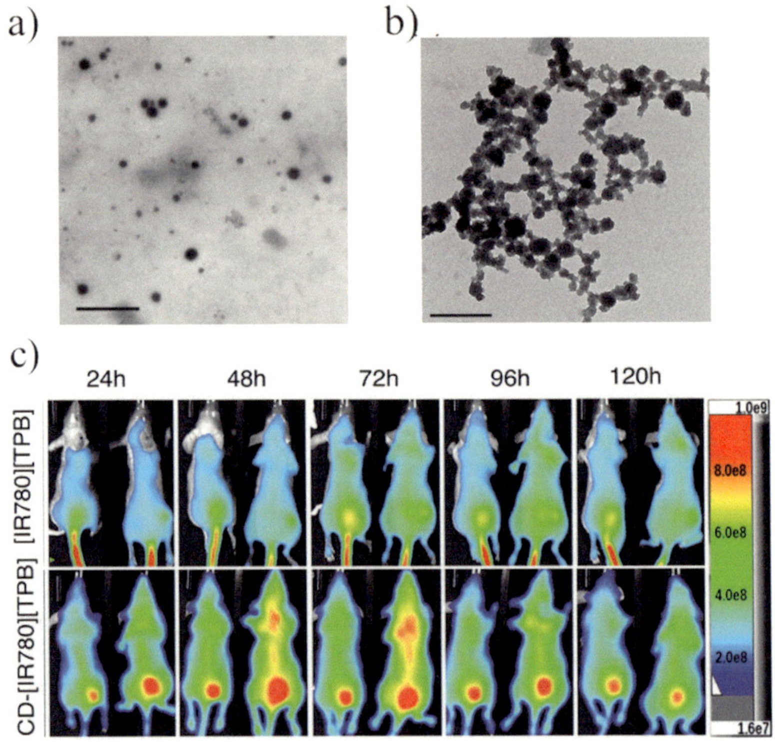

Figure 5.
TEM images of (a) [IR-780][TPB] nanoGUMBOS, (b) CD-[IR-780][TPB] nanoGUMBOS (scale bar represents 500 nm). (c) In vivo fluorescence of [IR-780][TPB] and CD-[IR-780][TPB] nanoGUMBOS at different time points.

CD-nanoGUMBOS using an MDA-MB-231 tumor xenograft model. Interestingly, NIR fluorescence intensity of mice demonstrated that CD-nanoGUMBOS distributed more rapidly and provided a higher tumor accumulation than nanoGUMBOS (**Figure 5c**). Moreover, these authors observed tumor decrease in mice treated with CD-[IR-780][TPB] nanoGUMBOS plus irradiation. Thus, these results indicated that CD-nanoGUMBOS have great potential as chemo-theranostic agents.

Broadwater et al. combined heptamethine cyanine cation ($[Cy]^+$) with several anions, such as iodide ($[I]^-$), hexafluoroantimonate ($[SbF6]^-$), and hexafluorophosphate ($[PF6]^-$), o-carborane ($[CB]^-$), along with bulkier anions such as tetrakis(4-fluorophenyl)borate ($[FPhB]^-$), cobalticarborane ($[CoCB]^-$), tetrakis (pentafluorophenyl) borate ($[TPFB]^-$), tetrakis[3,5-bis(trifluoro methyl)phenyl] borate ($[TFM]^-$), and Δ-tris(tetrachloro-1,2-benzene diolato) phosphate(V) ($[TRIS]^-$) to obtain several Cy-based organic salts [131]. Redox values, zeta potentials, HOMO energy level, as well as optical properties of all [Cy]-based organic salts were determined in that study. These results demonstrated that counterion exchange allowed tuning of HOMO energy levels of $[Cy]^+$. However, absorbance of all synthesized Cy-based organic salts spectra remained essentially the same.

The above cited authors then synthesized nanoparticles through a simple reprecipitation method. All Cy-based organic salts nanoparticles were determined to have an average size between 5–9 nm and were stable for 22 days. Following these experiments, Broadwater and collaborators evaluated cytotoxicity and phototoxicity of Cy-based organic salt nanoparticles against two different cell lines: human lung carcinoma (A549) and metastatic human melanoma (WM1158) in the absence and presence of 850 nm light. These results indicated that $[I]^-$, $[SbF6]^-$, and

$[PF6]^-$, $[CB]^-$ presented high cytotoxicity under both condition evaluated, making these compounds good candidates as chemotherapeutic agents. When $[Cy]^+$ was paired with $[FPhB]^-$ and $[CoCB]^-$ anions, nanoparticles of these compounds were determined to be slightly toxic at concentrations of 7.5 μM without NIR irradiation. However, when these compounds were irradiated with NIR laser, they were highly toxic at 5.5 μM concentrations, which indicates that [Cy][FPhB] and [Cy][CoCB] presented high potential as phototoxic agents. In contrast, compounds where $[Cy]^+$ was combined with bulky anions: $[TPFB]^-$, $[TFM]^-$, and $[TRIS]^-$ showed non-cytotoxicity against lung cancer cells. As a result, Broadwater, et al. concluded that these compounds could be employed as imaging agents. Based on cytotoxicity studies, the authors proved that toxicity of $[Cy]^+$ dye was tuned through counterion exchange. Finally, these authors evaluated *in vitro* imaging properties of all Cy-based organic salts nanoparticles. Analyses of these results demonstrated that non-toxic Cy-based organic salts could be employed at higher concentrations that resulted in a higher fluorescence intensity. Additionally, an *in vivo* evaluation of these compounds demonstrated that these compounds were accumulated in tumor sites.

2.2 Recycling of antimicrobial agents and other medicines

Antibiotics were introduced into modern medical practices in the 19th century with the discovery of sulfonamides in the 1930s and penicillin in the 1940s [132–134]. Without such medication, people often died from infections such as syphilis, gonorrhea, and pneumonia. Thus, use of these antibiotics represented the saving of many thousands of lives and a new era in medicine [135, 136]. Nevertheless, over the years since such discoveries, increased production, indiscriminate use, and over consumption of antibiotics has created an unfortunate outcome of antibiotic and multi-antibiotic resistant bacteria [137–139]. For this reason, the scientific community has begun to focus on syntheses of new antibiotics that could provide alternative therapies in order to avoid bacterial resistance mechanisms. However, syntheses of completely new antibiotics require great ingenuity, intense synthetic prowess, excellent purification, and considerable resources [140–142]. As an alternative to the foregoing strategy, several research groups have applied an ion metathesis strategy for antibiotic renewal and enhancement. Thus, recycling of current antibiotics have become a reality.

Florindo et al. synthesized ampicillin based ILs employing triethylammonium ($[TEA]^+$), choline ($[N_{1112}OH]^+$), trihexyltetradecylphosphium ($[P_{66614}]^+$), 1-ethyl-3-methylimidazolium $[C_2MIm]^+$, 1-hydroxy-ethyl-3-methylimidazolium $[C_2OHMIm]^+$ and cetylpyridinium $[C_{16}Pyr]^+$ as cations to tune crystalline forms and pharmaceutical properties [143]. In that study, water solubility, octanol/water partition coefficient ($K_{o/w}$), and phospholipid/water partition (K_p) of synthesized compounds were evaluated. Water solubility of active pharmaceutical ingredients (API) is of great importance because it determines the accessibility and distribution of API within the body. Using water solubility results at room temperature and at 37 °C, the following trend was observed: $[C_2OHMIM][Amp] > [N_{1112}OH][Amp] > [C_2MIM][Amp] > [TEA][Amp]$. All Amp-based ILs studied showed lower solubility as compared to [Na][Amp] at room temperature. However, cations with hydroxyl groups presented higher solubility at 37 °C than the parent compound. In contrast, Amp-ILs with longer carbon chains, such as $[P_{66614}]^+$ and $[C_{16}Pyr]^+$, showed K_p values higher than the parent compound, indicating that these compounds could interact better with cellular membranes. Based on results obtained in this study, $[N_{1112}OH][Amp]$ provided the most promising pharmaceutical properties with higher solubility, lower cytotoxicity, lower inflammation response and similar $K_{o/w}$ relative to the parent compound. Thus, these researchers confirmed that pharmaceutical properties from [Na][Amp] could be finely tuned through simple counterion exchanges.

In another study, the same researchers synthesized ciprofloxacin and norfloxacin fluoroquinolones (FQ) based protic ionic liquids (PILs) through reaction with mesylic acid ([Mes][H]), gluconic acid ([Glu][H]), and glycolic acid ([Gly][H]) to tune their crystalline forms and pharmaceutical properties to enhance their bioavailability [144]. In this case, similar properties as in previous studies were evaluated [143]. The authors observed a clearly increasing trend in aqueous solubility depending on the anion present in FQ-PILs: $[Gly]^- < [Mes]^- < [Glu]^-$. These observations were in agreement with $K_{o/w}$ studies obtained by these researchers. Similar K_p results for parent FQs were obtained for their respective organic salts, which indicated that interactions with cellular membranes were not affected. Based on these results, Florindo et al. concluded that FQ-based organic salts studied in this work presented high potential as alternatives to the original antibiotics.

Santos, et al. employed two FQs (ciprofloxacin and norfloxacin) to synthesize active pharmaceutical ingredient (API)-based ILs by combining their salts with the following cations: $[N_{1112}OH]^+$, $[C_{16}Pyr]^+$, 1-ethyl-3-methylimidazolium $[C_2MIm]^+$, 1-hydroxy-ethyl-3-methylimidazolium $[C_2OHMIM]^+$, 1-(2-hydroxyethyl)-2,3-dimethylimidazolium $[C_2OHDMIm]^+$, and 1-(2-methoxyethyl)-3-methylimidazolium ($[C_3OMIm]^+$) [145]. Water solubility of the synthesized compounds were evaluated, and the following trend was observed: $[EMIM]^+ < [Ch]^+ < [C_2OHDMIM]^+ \approx [C_3OMIM]^+ < [C_2OHMIM]^+$. This trend was similar to results reported in a previous study from the same group [145]. The authors determined IC_{50} concentrations of these FQ-ILs in this study against three bacteria: *Bacillus subtilis* (*B. subtilis*), *Staphylococcus aureus* (*S. aureus*) and *Klebsiella pneumoniae* (*K. pneumonia*) and compared results with IC_{50} values of the parent compounds. In order to evaluate if the synthesized compounds were more effective than the parent compounds, these researchers calculated the relative decrease of inhibitory concentration (RDIC) obtained by divided the minimum inhibitory concentration (MIC) values of each API-ILs by MIC of the corresponding API. Interestingly, most compounds evaluated in this work presented better antimicrobial activity than the original FQ with RDIC values higher than one [145].

Frizzo et al. employed sodium ibuprofen ([Na][Ibu]) and sodium docusate to synthesize API-based ILs [146]. Sodium cations in the parent compounds were replaced with ranitidine ($[Ran]^+$), diphenhydramine, glycine, or glycine ethyl cations. In this work, these researchers tested all synthesized compounds along with parent compounds against several types of bacteria and species of *Candidas* fungus. Evaluation of their results demonstrated that, in general, all API-ILs presented better antifungal activity than the precursors. For example, all ibuprofen-based ILs demonstrated antifungal activity. Interestingly, [Ran][Ibu] ILs presented antifungal activity when the parent compounds did not. In contrast, most API-ILs presented higher antibacterial activity than the parent compounds. These researchers synthesized ibuprofen- and docusate- based ILs that demonstrated high potential as antimicrobial and antifungal agents.

Ferraz et al. [147] also employed amoxicillin ($[seco\text{-}Amx]^-$) and penicillin G ($[seco\text{-}Pen]^-$) in combination with imidazolium, choline, ammonium, phosphonium and pyridinium cations to synthesize antibiotic based API-ILs. Resistant and sensitive Gram positive and Gram negative bacteria, including methicillin resistant *S. aureus* (MRSA ATCC 43300), were employed to test antimicrobial efficacy of these API-ILs through use of a broth micro dilution method. Evaluation of results obtained on sensitive strains demonstrated that only three of all evaluated compounds $[C_2OHMIM]$[seco-Amx], $[C_2OHMIM]$[seco-Pen], and [TEA][seco-Pen], produced lower or equal MIC values than the parent compounds and RDIC values equal to or higher than one. However, more interesting results were obtained when these compounds were tested against resistant bacteria such as *Escherichia coli* (*E. coli*) strains CTX M9 and CTX M2 as well as methicillin-resistant *S. aureus* ATCC 43,300. When tested against resistant *E. coli*, [C_{16}Pyr][seco-Amx] presented the highest RDIC value (> 100). In contrast, [C_{16}Pyr][seco-Amx] and [C_{16}Pyr][seco-Pen] were more effective against *S. aureus* ATCC 43,300 with RDICs higher than 1000 and 100, respectively. Another compound that was more effective than the commercial antibiotic was [N_{1112}OH][seco-Pen] with RDIC larger than 5. These findings clearly demonstrate that [C_{16}Pyr]$^+$ cation played an important role in antimicrobial activity of synthesized API-ILs acting in a synergetic way along with API present in the compound [147].

Cole and coworkers proposed recycling of antibiotics into GUMBOS [15]. In this work, ampicillin based GUMBOS (Amp-GUMBOS) were synthesized through a simple metathesis reaction where a sodium cation was replaced by hexadecyl-methyl-imidazolium ([C_{16}MIm]$^+$), hexadecyl-dimethyl-imidazolium ([$C_{16}M_2$Im]$^+$) and [C_{16}Pyr]$^+$. These Amp-ILs were tested against Gram negative and positive bacteria and compared to parent compounds. Interestingly, MIC values obtained for Amp-GUMBOS in these experiments demonstrated that these concentrations were between 2 to 43 times lower than MIC values determined for ampicillin.

Following their previous studies, Cole and co-workers employed chlorhexidine and ampicillin to synthesize antibacterial GUMBOS [14]. These two antibacterial agents are commonly used in veterinary practices to treat and/or prevent the presence of *E. coli* strains that are commonly found in cattle. The presence of *E. coli* strains could produce severe illness in humans, especially *E. coli* O157:H7, which is well-known because it produces bloody diarrhea, haemorrhagic colitis and haemolytic uraemic syndrome in humans [148–150]. For this reason, development of a prophylactic treatment for this type of bacteria are highly desirable for eradication or minimization in cattle to prevent human illnesses. Cole et al. evaluated the efficacy of this Chlorhexidine di-ampicillin GUMBOS to kill several *E. coli* O157:H7 strains isolated from different sources such as chicken, pork, beef, apple cider, burger and humans [14]. Chlorhexidine di-ampicillin GUMBOS was found to kill *E. coli* strains more effectively since these GUMBOS presented MIC values much lower than the parent compounds and their unreacted stoichiometric mixture. Moreover, interaction indices indicated that this antimicrobial GUMBOS presented a synergetic mechanism effect. Chlorhexidine is a commonly used antiseptic; however, it presents high cytotoxicity against normal cells. For this reason, these researchers studied cytotoxicity of GUMBOS, parent compounds and unreacted stoichiometric mixtures in Hela cells. Interestingly, GUMBOS were less toxic than parent chlorhexidine, reaching 93% cell viability.

In another work, Cole and coworkers recycled four β-lactam antibiotics (ampicillin, cephalothin, carbenicillin and oxicilin) into GUMBOS, by combining them with chlorhexidine diacetate [151]. Twenty-five bacteria isolates were obtained from several sources, where most of these were resistant or multi-resistant to antibiotics. These four β-lactam – based chlorhexidine GUMBOS were tested against these isolates. Results obtained by Cole and coworkers demonstrated that these β-lactam – based chlorhexidine GUMBOS were more effective against these isolates with MIC values in a range between 0.1 to 32 μM as compared to parent compounds with higher MIC (5 to >1250 μM). Moreover, in this report Cole et al. evaluated if these GUMBOS presented a synergetic, additive or antagonist effect relative to their unreacted mixtures of stoichiometric equivalents. Interestingly, these researchers found that for most GUMBOS studied, the observed effect was synergetic [152].

Neisseria gonorrhoeae is another bacterial target that is primarily sexually transmitted and responsible for the disease gonorrhea [153]. In recent years,

N. gonorrhoeae resistance to current treatments have been isolated and reported around the world [154, 155]. Lopez et al. have synthesized GUMBOS from an antiseptic octenidine and a discontinued antibiotic carbenicillin as a possible alternative to reduce and minimize *N. gonorrhoeae* transmission [156]. The zone of inhibition (ZOI) for *N. gonorrhoeae* strains and clinical isolates were studied in the presence of GUMBOS, parent compounds, and antibiotics currently employed for gonorrhea treatment. Evaluation of results obtained demonstrated that synthesized GUMBOS presented an additive effect as compared to the parent compounds as well as an equivalent antimicrobial activity like azithromycin.

3. Sensing materials

Sensing strategies for a variety of systems, from biological targets [157], environmental and regulatory applications [158, 159], mechanical integrity of structures [160–162], and more [60, 163], are continuously under investigation in the scientific community. In general, recognition can be categorized into two different methodologies: targeted and non-targeted [164]. Targeted strategies require materials that are designed to respond to specific analyte(s) and thus, require a high degree of specificity for singular analytes [63, 159, 165]. Differential strategies, however, can potentially provide information within convoluted and complex mixtures based on several non-specific sensors or one sensor with multi-layered responses to different analytes [166]. In the following sections, solid-state ionic materials for various sensing applications are discussed.

3.1 Ratiometric sensing: fluorescence imaging

Previous investigations using fluorescent imaging with solid-state ionic materials have undergone scrutiny to prevent or reduce self-quenching between dye molecules in order to enhance properties such as excitation energy transfer and achieve on/off switching in nanoparticle structures [167–172]. Traditionally, dye self-quenching has been rectified by introducing bulky side-chains into the molecular structure via synthetic organic chemistry [172–174]. However, this type of strategy requires several synthetic and purification steps that result in increased expense. In contrast, large counterions were observed to also inhibit this self-quenching phenomenon in a much more facile manner through a simple ion metathesis reaction [28, 50]. Several research groups have capitalized on this strategy to study polymeric nanoparticle encapsulated rhodamine-derived GUMBOS, respective photophysical properties, and cellular uptake ability for imaging applications along with targeting agents to provide organelle contrast [28, 172, 175, 176].

More recently, researchers have diversified beyond cellular imaging techniques. For example, Severi et al. have explored polymer encapsulation of nanoprobes that undergo efficient Förster resonance energy transfer (FRET) for potential point-of-care applications with smartphones [177]. In this study, ester-modified cations rhodamine 110 and 6G cations ($[R110]^+$ and $[R6G]^+$, respectively) were employed as FRET donor dyes with bulky tetrakis[3,5-bis(1,1,1,3,3,3-hexafluoro-2-methoxy-2-propyl)phenyl]borate trihydrate ($[F12]^-$) and tetrakis(perfluoro-tertbutoxy) aluminate ($[F9\text{-}Al^-]$) counterions [176]. These Ion pairs were encapsulated with DNA cancer marker (survivin) targeted polymer nanoparticles, which were also functionalized using a red-emitting oligonucleotide-functionalized dye as a FRET acceptor. After nanoparticle size, quantum yield (QY), FRET acceptor concentration optimization, and evaluation of FRET capabilities, encapsulated [R6G][F9-Al] nanoprobes were evaluated for red, green, blue (RGB) survivin DNA marker detection in solution using fluorescence spectroscopy. These researchers found that their designed system had a limit of detection of 3pM. Upon optimization of microscopic and digital imaging, these researchers also found that using an iPhone SE, [R6G][F9-Al] as an encapsulated FRET donor in their designed nanoprobe allowed a 10pM limit of detection. Thus, these researchers demonstrated that [R6G][F9-Al] was successfully employed as a visualization agent for potential development of a point of care ratiometric imaging method.

In another study, McNeel et al. expanded upon counterion metathesis by synthesizing a strategic three-component nanoGUMBOS compound for selective imaging of breast cancer cells [27]. Two of three components selected were dianionic fluorescein ($[FL]^{2-}$) and cationic rhodamine B ($[RhB]^{+}$), which could undergo pH-dependent FRET [178]. These researchers approached their triple-GUMBOS synthetic design through pH manipulation with $[FL]^{2-}$, rhodamine B chloride [RhB][Cl], and $[P_{66614}]$[Cl] as a hydrophobic agent, to yield $[P_{66614}]$[RhB][FL] triple GUMBOS. The resultant compound was then employed for nanoGUMBOS synthesis, and when precipitated from water with neutral pH, nanoparticles of approximately 4.4 nm ± 0.7 nm were obtained. However, when using other pH values for nanoGUMBOS synthesis, these researchers determined that nanoGUMBOS sizes and size distributions varied. Absorbance and fluorescence emission properties from low to high pH values were reported, and noticeable ratiometric changes in spectra were observed. A linear ratiometric trend corresponding to pH-dependent FRET responses was observed between moderate pH values (approximately pH 5.0 to 7.0), from which quantitative information may be derived. To further demonstrate the applicability of this three-component nanoGUMBOS system, these investigators also conducted fluorescence microscopy imaging studies with normal and cancerous breast cells. These studies demonstrated that nanoGUMBOS maintained clear selectivity for breast cancer cells since, as cells were illuminated. In contrast, normal cells remained dim. Therefore, three-component nanoGUMBOS were determined useful for both pH sensing and fluorescence imaging of breast cancer cells without the use of polymer encapsulation [27].

Another application for FRET-based sensing of solid-state ionic materials is described by Ashokkumar et al. [179]. In this work, oxygen sensing nanoparticle probes for cellular systems were developed using a polymer encapsulated novel cyanine dye called $[BlueCy]^{+}$ tetrakis(pentafluorophenyl) borate ($[F5\text{-}TPB]^{-}$) or [BlueCy][F5-TPB] that was also loaded with oxygen sensing platinum octaethylporphyrin (PtOEP) as a FRET acceptor. In this case, $[BlueCy]^{+}$ was designed as FRET donor and synthesized from two cyanine dyes: 2-methyl-3-octadecylbenzo[d]thiazol-3-ium iodide and 3-methyl-2-(methylthio)benzo[d]thiazol-3-ium iodide. After dye encapsulation into poly(methyl methacrylate-co-methacrylic acid) and poly(lactic-co-glycolic acid), PMMA-MA and PLGA, respectively, it was determined that both dyes were successfully incorporated into PMMA-MA. Nanoparticle sizes, PtOEP loading, and photophysical properties were evaluated. These investigators determined that dye encapsulated PMMA-MA nanoparticles were 40 nm in diameter with 17% QY after reprecipitation from dioxane. Moreover, after testing several ratios of donor dye loadings, ratios of 1:100 (PtOEP:[BlueCy][F6-TPB]) demonstrated good FRET efficiency. Solution based experiments for oxygen sensing were performed, and ratiometric trends were demonstrated for oxygen rich and poor environments. After confirming low phototoxicity when incubated with HeLa cells, the investigators conducted further studies with FRET nanoparticles in low and normal oxygen environments. HeLa cells were incubated with FRET nanoparticles in a microfluidic device, and an oxygen gradient was introduced by application of an oxygen scavenger. Resultant emission gradients were observed after fluorescent microscopic images were obtained. Ultimately, these researchers demonstrated the utility of their nanoprobe for detection of cancer cells via microfluidic application. As a result, the authors concluded that this probe could also be used to visualize oxygen gradients in cancerous cells.

In another study from the Warner research group, nanoGUMBOS were synthesized and evaluated as ratiometric sensors for reactive oxygen species (ROS) [42]. Cong et al. designed binary nanoGUMBOS using reprecipitation of 1,1′-diethyl-2,2′-cyanine and 1,1′-diethyl-2,2′-carbocyanine bis(perfluoroethylsulfonyl) imide ([PIC][NTf_2] and [PC][NTf_2], respectively). Optimal FRET efficiency was determined to be 10:1 [PIC]:[PC] molar ratio, and binary nanoparticles shapes were classified as nanodiamonds with spectrally consistent J-aggregation. Analysis of variance (ANOVA) was employed to investigate reactivity of ROS with nanoGUMBOS. Significant differences were observed for hydroxyl radical ($^{\bullet}OH$) over four other evaluated ROS, indicating selectivity of this binary nanoGUMBOS system toward $^{\bullet}OH$ species. Moreover, these investigators observed a linear trend for ratiometric sensing of this probe at various concentrations of $^{\bullet}OH$ in the presence of singlet oxygen ($^1O_2$). Further, potential applications in imaging were investigated, nanoGUMBOS were incubated with breast cancer cells and exposed to oxidative stress. Fluorescence emission changes before and after oxidative stress indicated results in agreement with solution-based studies. Therefore, a binary ratiometric nanoGUMBOS probe was developed for potential quantitative ROS imaging studies using a facile method.

3.2 Differential sensing: biological applications

Biosensing of mixtures of biomarkers and/or proteins is of particular interest for disease diagnosis and treatment [180–182]. Many current methods, such as enzyme-linked immunosorbent assay (ELISA) or polyacrylamide gel electrophoresis (PAGE) coupled to mass spectrometry, require expensive resources and labor intensive steps [182–184]. Organic salts are of increasing interest for development of fluorescent sensor arrays for protein detection and discrimination as they are easily tunable for increasing hydrophobicity, traditionally more stable upon ion exchange, and require little resources for purification [90].

Galpothdeniya and coworkers used partially selective 6-(p-toluidino)-2-naphthalenesulfonate sodium salt ([TNS][Na]) in an ion exchange metathesis reaction with cations tetrabutylphosphonium ([P_{4444}]$^+$), benzyltriphenylphosphonium ([BTP]$^+$), 4-nitrobenzyltriphenylphosphonium ([4NBP]$^+$), and tetraphenylphosphonium ([TPP]$^+$) in order to obtain four different GUMBOS [59]. These investigators rationalized that, as a result of partial selectivity to hydrophobic regions of proteins, TNS-based GUMBOS would make facile, suitable candidates to generate a sensor array for proteins. Proteins such as human serum albumin (HSA), fibrinogen, α-antitrypsin (α-Ant), immunoglobulin G (IgG), β-lactoglobulin (β-Lac), ribonuclease A (RNaseA), α-chymotrypsin (α-CTP), transferrin (Trans), lysozyme (Lys) were used for sensor array development. Sensor responses were collected at various concentrations of proteins.

As a result of notably larger sensor responses, [TNS]-based GUMBOS were determined to have highest sensitivity to HSA, α-Ant, and β-Lac proteins. For this reason, the investigators employed responses for sensor responses to different concentrations of HSA, α-Ant, and β-Lac for multivariate analysis. Both sensor response values and corresponding protein concentrations were employed to build a principal component analysis (PCA) model. By employing the first two principal components (PCs), which accounted for 99.72% of the variance, a linear discriminant analysis (LDA) model with cross-validation was constructed reaching 100% discrimination accuracy. These researchers noted that the highest sensor responses were obtained for HSA and α-Ant. Thus, these sensor responses were employed to generate anther PCA model in order to evaluate discrimination between these two proteins regardless of protein concentration. In this model, the first two PCs accounted for 99.91% variance, and LDA with cross-validation resulted in 91.7% accuracy. To improve this accuracy, these investigators normalized sensor responses for each protein, constructed a PCA model with the first three PCs corresponding to 98.29% variance. These three PCs were employed for LDA construction and, with cross-validation, accuracy resulted in 100% discrimination. Furthermore, five mixtures of different HSA: α-Ant ratios were evaluated for mixture discrimination analysis. In this case, PCA followed by LDA resulted in 100% discrimination accuracy. Thus, TNS-GUMBOS were evaluated and confirmed as useful materials for protein sensor arrays for analyses of serum proteins HSA, α-Ant, and β-Lac.

More recently, Pérez and coworkers developed a nanoGUMBOS sensor array based on three fluorescent thiacarbocyanine ($[TC0]^+$, $[TC1]^+$, and $[TC2]^+$) dyes with two anions ($[BETI]^-$ and $[NTf_2]^-$) for discrimination of several proteins [182]. NanoGUMBOS and microGUMBOS of these six compounds varied in size and shape, from circular shapes with $[TC0][NTf_2]$ and sizes around 25 nm, to [TC1][BETI] with rod-like shapes and an average size of 1.2 ± 0.5 μm by 0.21 ± 0.08 μm, and $[TC2][NTf_2]$ displayed triangular profiles with average dimensions 200 ± 10 nm by 177 ± 80 nm. Aggregates of nanoGUMBOS of [TC0]- and [TC2]-GUMBOS exhibited absorbance spectral characteristics representative of H-aggregation, while [TC1][NTf2] and [TC1][BETI] both resulted in spectral peaks representative of J-aggregation.

In the above study, seven proteins were investigated, including the four most abundant serum proteins: HSA, IgG, transferrin (Trans), and fibrinogen (Fib), along with three non-serum proteins hemoglobin (Hb), cytochrome C (CytC), and lysozyme (Lys), with each protein exhibiting different physical characteristics. The investigators observed different response patterns for each protein. In this work, these researchers determined that employing raw data was optimal for constructing an LDA model, in which 100% discrimination accuracy of proteins was achieved. Among different protein concentrations, sensor responses were determined to be stable between 0.1 to 20 μg/mL. Mixtures of two proteins, HSA and Hb, were also investigated in this work. Various weight ratios of HAS:Hb mixtures from 100% HSA to 100% Hb, were evaluated and 100% accuracy was achieved when LDA was constructed using these response patterns. However, 80:20 HSA:Hb was observed to be an outlier with the lowest canonical score values, and further analysis using hierarchical cluster analysis determined this dataset to be less related to other ratios. Protein spiked artificial urine with 5 μg/mL protein concentration was employed to evaluate sensor array performance in real samples, and LDA model performance achieved 100% discrimination accuracy. Thus, a series of TC-based GUMBOS were successfully synthesized into nanoGUMBOS and microGUMBOS and developed as protein sensor arrays capable of 100% discrimination in complex mixtures.

3.3 Differential sensing: volatile organic compounds

ILs have been explored for quartz crystal microbalance (QCM) applications as chemosensors for detection and discrimination of volatile organic compounds (VOC) [185–188]. However, for these investigations, differentiation of VOCs sensor response relied on concentration and molecular composition of an analyte. In 2012, Regmi and coworkers developed a system to correlate sensor responses using GUMBOS-polymer composite [75]. In this work, investigators characterized and explored the responses of cellulose acetate and 1-butyl-2,3-dimethylimidazolium hexafluorophosphate (CA-[BM_2IM][PF_6]). By carefully evaluating characteristic responses upon exposure to control sensors, composite material, and confirming results using molecular dynamic simulations, investigators determined that sensor response recorded as changes in frequency were directly proportional to changes in motional resistance. Thus, these researchers successfully derived molecular weight trends from their composite sensor.

Another exploration demonstrated that counterion exchange using only GUMBOS coatings on quartz crystal resonators (QCRs) could provide VOC differentiation. In 2015, Regmi et al. explored trihexyltetradecylphosphium copper phthalocyanine-3,4′,4″,4‴-tetrasulfonic acid ([P_{66614}]$_4$[$CuPcS_4$]) and trihexyltetradecylphosphium copper(II) meso-tetra(4-carboxyphenyl)porphyrin ([P_{66614}]$_4$[CuTCPP]) as sensing materials [73]. Each GUMBOS sensor successfully allowed detection of a variety of VOCs, such as acetone, acetonitrile, nitromethane, toluene, chloroform ($CHCl_3$), methanol (MeOH), ethanol (EtOH), 2-propanol, 1-propanol, 1-butanol, and 3-methyl-1-butanol. Both sensor responses readily allowed detection of multiple alcohols at relatively low detection limits when compared to other polar and nonpolar analytes. When compared to IL trihexyltetradecylphosphium bis(trifluoromethanesulfonimide), [P_{66614}]$_4$[CuTCPP] provided higher frequency response signals upon exposure to MeOH vapor, and more rapidly achieved baseline with efficient replicate results. Thus, these investigators demonstrated that use of copper(II) porphyrin counterion in GUMBOS allowed investigators to achieve high selectivity in sensor responses to VOCs [73]. Since these reports, there have been other explorations into IL and/or polymer-IL composite responses for VOC detection, and many have attained discrimination via statistical techniques to access virtual and multi-sensor arrays [69–72, 189].

Since VOCs are frequently found as complex mixtures, Vaughan et al. have proposed development of a multi-sensor array (MSA) employing copper(II) phthalocyanine or [$CuPcS_4$]-based GUMBOS sensors [68]. An example of such a sensory coating scheme is shown in **Figure 6**. VOCs studied represent compounds from different classes, such as dichloromethane (DCM), MeOH, 1-propanol, toluene, $CHCl_3$, heptane, hexane, and benzene. In this work, [P_{4444}]$^+$, tributyl-n-octylphosphonium ([P_{4448}]$^+$), tetrabutylammonium ([TBA]$^+$), 3-(dodecyldimethyl-ammonio)propanesulfonate ([DDMA]$^+$) were employed as cations for [$CuPcS_4$]$^{4-}$ to generate four different sensory coatings. Each coating displayed different layering characteristics as determined by SEM. Upon exposure to VOCs, each sensor presented analyte specific response patterns. Using original data, and quadratic discriminant analysis (QDA) with cross-validation, the resultant accuracy was determined to be 98.6%.

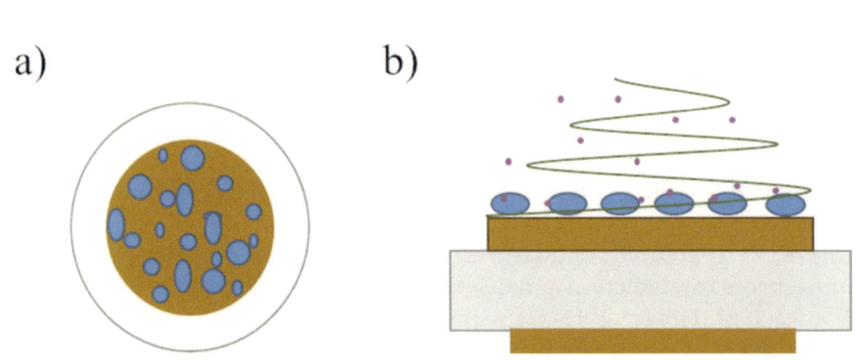

Figure 6.
(a) Representative example of GUMBOS coated QCR; (b) analyte sensing and harmonic wave pattern of QCR on electrode surface.

Thus, [$CuPcS_4$]-based GUMBOS responses were successfully employed to build a VOC-MSA to achieve high accuracy discrimination [68].

4. Optoelectronic developments

With increasing global commercialization of state-of-the-art optoelectronic displays, the demand for higher performance and flexible materials has also increased [107–110]. In general, these devices are comprised of emissive layers between electrodes along with several other electronically active layers. Organic light emitting diodes (OLEDs) and organic photovoltaics (OPVs) have been the central target for a multitude of research groups, from organic emissive layer development to full device performance [107–111]. Counterion strategies using cations and anions within active layers for enhancement on optoelectronic device fabrication to effects on emission and device function will be discussed in the following sections.

4.1 Counterion strategies for light emissive layers in OLEDs

Scientists have optimized several characteristics for targeted OLED development where they require consistent uniformity of emissive layers for potential manufacturing production [190], low crystallinity to prevent non-linear optical activity [191, 192], resistance to oxidation and water [193], and high thermal stability and optical purity [194]. Ionic transition metal complexes (ITMCs) are of huge interest as a wide range of emissive hues is easily achievable, synthesis is relatively simple, and they have desirable luminescent properties [195]. However, traditional methods of OLEDs fabrication involves vacuum evaporation deposition, or vacuum thermal evaporation (VTE) [196], which involves uniform coating of emissive layers [197].

In this regard, Dongxin Ma and coworkers have investigated four cationic iridium complexes as candidates for VTE through counterion control [195]. By incorporating large non-coordinating anions, these investigators achieved VTE iridium-based ionic emissive layers. The anions [PF_6]$^-$, [TPFB]$^-$, and tetrakis[3,5-bis(trifluoromethyl)phenyl]borate ([BArF]$^-$) were employed for quantum chemical calculations, and the investigators determined that distances between iridium and boron atoms were larger than 8 Å with both [TPFB]$^-$ and [$BArF_{24}$]$^-$ counterions. In comparison, distances between iridium and phosphorous was determined to be 6 Å, as a result of a larger partial positive charge on the phosphorous atom. Compounds synthesized from metathesis with bulkier anions [TPFB]$^-$ and [$BArF_{24}$]$^-$ were employed for device fabrication using VTE as larger interatomic distances were presumed more suitable for phase transition, and device performance was evaluated. These investigators found devices ranged from blue to red-orange with external quantum efficiencies (EQEs) ranging from 1.2% (blue emission) to 8.1% (yellow emission) [195]. In 2018, these anions were also employed to produce two red-orange devices based on cationic iridium compounds, and these compounds were useful as dopants in 4,4′,4″-tris(carbazol-9-yl)triphenylamine emissive layers (TCTA) to produce white OLEDs with Commission Internationale de L'Eclairage (CIE) coordinate values equal to (0.33, 0.34), that were near to the required values (0.33, 0.33) [198].

More recently, Bai and coworkers investigated counterion-tuning strategies for a sky-blue fluorescent Ir-cation for VTE [199]. Instead of boron-based anions to improve VTE, the investigators strategically focused on bulky sulfonate-containing anions that also contained electron-deficient oxadiazole and triazine structures, such as 3,5-bis(5-(4-(tert-butyl)phenyl)-1,3,4-oxadiazol-2-yl)benzenesulfonate ([OXD-7-SO_3]$^-$), 4-(4,6-diphenyl-1,3,5-triazin-2-yl)benzenesulfonate

([TRZ-p-SO_3]$^-$), and 3-(4,6-diphenyl-1,3,5-triazin-2-yl)benzenesulfonate ([TRZ-m-SO_3]$^-$). These structures were expected to not only provide VTE capabilities, but also improve carrier transport and trapping efficiencies that would improve overall efficiency and blue-emission of OLEDs. These researchers concluded that devices fabricated with [TRZ-m-SO_3]$^-$ and [TRZ-p-SO_3]$^-$ anions resulted in better overall device performances, slightly decreased CIE x-coordinate value, and displayed the largest external quantum efficiencies (EQEs) of 12.3 and 12.4%, respectively [199].

Carbazole-containing compounds with expanded conjugation are known to provide efficient blue emission, although they often require high labor and resource costs. In this regard, Siraj et al. synthesized carbazole imidazolium iodide ([CI][I]) along with analogues containing [OTf]$^-$, [NTf_2]$^-$, and [BETI]$^-$ anions as respective GUMBOS in an efficient manner [54]. These GUMBOS were then compared to parent [CI][I] to evaluate counterion effects on thermal and photochemical properties that relate to performance for blue-emitters for OLEDs. In this study, non-uniform packing was observed in all GUMBOS as a result of cation structure. All ion-exchanged GUMBOS also demonstrated significantly higher thermal stabilities with onset degradation temperatures ranging from 310 to 417 °C as determined by thermal gravimetric analysis (TGA), where increasing size of anion yielded increased degradation temperature ([I] < [OTf]$^-$ < [NTf_2]$^-$ < [BETI]$^-$). Similarly, QYs were increased with ion exchanges. In methanolic solution, [CI][BETI] was determined to have the largest QY of 99%, followed by [CI][OTf] with 94%, [CI][NTf_2] with 73%, and [CI][I] with 25%. Thus, this demonstrated that hydrophobic counterion exchange affects photophysical properties of the CI-cation.

While VTE has dominated OLED manufacturing, it often requires expensive equipment and is both energy and time consuming [200]. For this reason, several researchers have explored solution processing methods, such as spin coating [200], electrospray deposition [54], along with other methods [201, 202] to provide faster, more inexpensive fabrication procedures [203]. In this report, [CI][OTf], [CI][NTf2], and [CI][BETI] GUMBOS were used to fabricate thin films on quartz glass with electrospray deposition [204], and uniform coating was achieved and confirmed by scanning electron microscopy and fluorescence microscopic analysis. Solid-state emission spectra displayed very slight red-shifting from methanolic spectra of ion-metathesis GUMBOS. In addition, photostabilities were investigated, and [CI][BETI] displayed an irradiation-induced increase in photostability. In contrast, [CI][OTf] and [CI][NTf_2] were relatively stable while irradiated for 3000 s. Moreover, cyclic voltammetry and quantum chemical calculations further supported spectral properties of evaluated CI-based GUMBOS.

In 2016, Zhang and coworkers designed a novel cyanopyridinium stilbene cation ([Py]$^+$) in order to examine the influence of counterion effects on solid-state photophysical properties [205]. Chloride ([Cl]$^-$), nitrate ([NO_3]$^-$), tosylate ([OTs]$^-$), and [TPB]$^-$ anions were employed in this study to form Ion pairs, and the resultant compounds showed little fluorescence in solution. When explored as films, blue-shifting of emission peaks occurred and increased with increasing hydrophobicity of counterions; QYs also increased following this trend. However, [Py][TPB] GUMBOS were non-emissive in solid-state. In order to understand this variance in trend, investigators used X-ray crystallography and quantum chemical calculations. From these studies, the authors determined that dimeric fluorophore aggregates were responsible for emission in GUMBOS. In [Py][TPB], fluorophores became dilute as a result of bulky anions, resulting in very weak fluorescence emission. This was also confirmed in quantum chemical studies, where intramolecular charge transfer characteristics were confirmed through prediction of frontier molecular orbital placement to reveal donor-σ-acceptor properties for dimeric stacking.

In another study, expansion of applications of the propidium dication ($[P]^{2+}$) was investigated by exchanging iodide counterions for $[OTf]^-$, $[NTf_2]^-$, and $[BETI]^-$ anions to generate P-based GUMBOS for potential solid-state applications [53]. Thermal, spectral, photo-physical, computational, and electrochemical properties were investigated for all P- based GUMBOS. While [P][OTF] retained physical properties similar to parent dye, such as solubility in more polar solvents, thermal degradation, and higher relative crystallinity, similar to the parent compound. In contrast, $[P][NTf_2]$ and [P][BETI] GUMBOS were more soluble in hydrophobic solvents, more amorphous, and displayed higher thermal stability. A trend was observed where increasing solvent hydrophobicity increased fluorescence lifetime and QY values. The highest fluorescence lifetime and QY for [P][BETI], followed by $[P][NTf_2]$, was observed in DCM. In this regard, the authors proposed that this effect may be a result of hydrophobic counterion stabilization of excited-state $[P]^{2+}$, a phenomenon that resembles the original sensing behavior of the parent compound [P][I] [206]. These investigators also performed cyclic voltammetry to determine oxidation and reduction potentials for each P-based GUMBOS, as well as solution-phase QY calculations. It was determined through computational experiments that electronic transitions would lead to an increased propensity for torsional twisting in the solid state [207]. From these studies, the authors concluded that by simple counterion exchange, applications for propidium dication may be expanded beyond biological probes to potential candidates in optoelectronic devices [53].

4.2 Recent advances in dye-sensitized solar cells

Dye-sensitized solar cells (DSSCs) are an emerging next-generation technology in OPVs [208]. Through intrinsic characteristics such as natural transparency, good efficiency in low light conditions, flexible substrate production and more, applications may be expanded to windows, indoor fixtures, and wearable electronics [209, 210]. Incorporation of two or more complementary sensitizing dyes allows for potential absorption of all wavelengths of sunlight to achieve much higher power conversion efficiencies (PCEs) [208, 211–214]. When implemented in this field, tunable investigations of co-sensitizing dyes that are also ionic and primarily limited to structural variations of zwitterionic squaraine-heptamethine structures rather than ion pairs [215, 216]. Polymethine, or cyanine dyes, however, have been extensively studied in OPV technologies, and several groups have begun explorations into counterion application in DSSCs [217–222].

In 2012, Jordan et al. from the Warner research group reported synthesis and characterization of $[PIC][NTf_2]$ and [PIC][BETI], along with fabrication of respective nanoGUMBOS [36]. Optical properties were compared to the parent [PIC][I], and nanoGUMBOS were synthesized and characterized using TEM and scanning electron microscopy (SEM). Optical properties of the resultant PIC-based nanoGUMBOS were also investigated. These investigators concluded that $[PIC][NTf_2]$ nanodiamonds resulted in a significant increase in fluorescence emission intensity, which could be a result of J-aggregation. The authors hypothesized that [PIC][BETI] nanorods from H-aggregates only slightly increased fluorescence intensity. In 2014, Sarkar and coworkers investigated morphology, size, and current–voltage characteristics of these PIC-based nanoGUMBOS using atomic force microscopy (AFM) and conductive probe-AFM (CP-AFM) [56]. Results from CP-AFM indicated that when the voltage was swept between 1 and − 1 Volts, current values within the range of approximately 10^{-7} to 10^{-8} Amps could be achieved. Raman spectroscopy was employed to monitor anion effects on aggregation changes via changes in intensity. These researchers confirmed that $[PIC][NTf_2]$

nanoGUMBOS exhibited J-aggregation while H-aggregation was observed in [PIC] [BETI] nanoGUMBOS. Thus, researchers from both investigations showcased anion dependent nanoparticle morphology and respective effects on spectral and electrochemical properties of broadly absorbing PIC-based nanoGUMBOS that had potential uses in DSSCs.

Kolic et al. have investigated different GUMBOS, including the aforementioned PIC-based GUMBOS, to determine effects on DSSC performances [34]. These dyes were employed as energy relay dyes (ERDs) in electrolyte solutions, where FRET occurs to donate electrons from ERD molecules in electrolyte solution to photosensitizing dye at the electrode surface. **Figure 7** represents a proposed scheme for electron transfer processes involving GUMBOS-ERDs DSSCs. Photoactive dyes such as rhodamine B ($[RhB]^+$), $[PIC]^+$, thiacarbocyanine ($[TC1]^+$), and tetracarboxyphenylporphine ($[TCPP]^{4-}$) precursors underwent ion exchange with appropriate counterions, such as $[NTf_2]^-$, $[BETI]^-$, and $[P_{66614}]^+$, and were further evaluated for counterion effects on ERD performance. Among various GUMBOS studies, investigators determined that $[RhB][NTf_2]$ and $[P_{66614}]_4[TCPP]$ GUMBOS yielded most promising PCEs devices. These investigators hypothesized that this was a result of inherent high molar extinction coefficients and QYs for these respective compounds. They also noted that devices employing $[NTf_2]^-$ anions resulted in higher respective device efficiencies than those of the parent dye or $[BETI]^-$ anions. One deviation of this trend, however, was the case of [TC1][TPB], which demonstrated a much higher QY. Overall, these authors were able to elucidate anion trends for GUMBOS-ERDs and confirm their utility as FRET cosensitizing agents in DSSCs [34].

Other works have recently focused on incorporating metal-based GUMBOS as redox shuttles for sensitizer regenerating agents in DSSCs as well. In 2016, Huckaba and coworkers employed a cobalt(II/III) redox shuttle ($[Co(bpy)_3]^{2+/3+}$) with $[NTf_2]^-$ as a non-coordinating anion with indolizine sensitizers [223]. Device PCEs ranged from 3.04 to 8.10% efficiencies, which were comparable to employing common redox shuttle, iodide/triiodide (I^-/I_3^-) (3.74–7.99%). Additionally, a copper(I/II) redox shuttle ($[Cu(tmby)_2]^{+/2+}$) with $[NTf_2]^-$ counterion was recently employed with indoline derivatives as sensitizers [224]. This study determined that this redox system rapidly regenerated indoline dyes within the range of tens of nanoseconds, several orders of magnitude faster than cobalt(II/III) shuttle $[Co(bpy)_3]^{2+/3+}[NTf_2]_{3/2}$ [224]. As a result of the volatility of the organic solvent employed in electrolyte solutions, some groups have expanded investigations into non-volatile routes for DSSC fabrication [211, 225, 226]. Cao and coworkers have developed a solid-state DSSC (ssDSSC) based on $[Cu(tmby)_2]^{+/2+}[NTf_2]_2$ redox shuttle for a hole transport layer [227]. In comparison to other copper redox shuttles, PCEs were

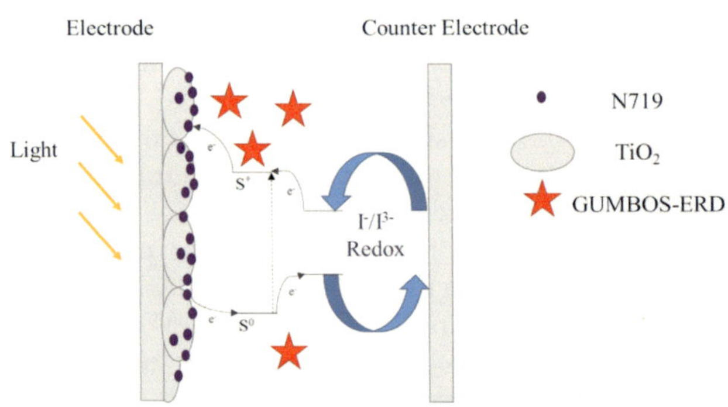

Figure 7.
Schematic of DSSCs fabricated with GUMBOS-ERDs and potential operational mechanism [34, 208].

determined to be much higher with this novel ssDSSC at 11% versus 4.5 or 2%, respectively. Thus, using their trilayered approach with co-sensitizer (Y123) and their solid-state hole transport material, scientists successfully demonstrated charge separation in a novel ssDSSC.

4.3 Organic salts to reduce work function in optoelectronics

Work function (WF) is a metric by which charge transfer at electrodes is measured to determine electron injection efficiencies of optoelectronics [228]. For this reason, many groups have targeted improving device efficiency by optimizing charge transport at interfacial layers by including electroactive coatings[228–230]. Incorporating ethoxylated polyethylenimine (PEIE) at electrode interfaces of OLEDs was previously demonstrated by Zhou and coworkers to reduce WF [231]. In 2019, Ohisa et al. hypothesized that incorporation of tetraalkylammonium salts ([TRA][X]) into PEIE layers could further reduce required WFs for OLEDs [231, 232]. Alkyl chain lengths that varied between tetraethyl ($[TEA]^+$), tetrabutyl ($[TBA]^+$), and tetrahexyl ($[THA]^+$) ammonium groups were studied in this report. Different anions were employed, and a series of salts were investigated for each ammonium cations. Anions employed ranged from $[Cl]^-$, bromine ($[Br]^-$), $[I]^-$, acetyl ($[Ac]^-$), thiocyanate ($[SCN]^-$), or tetrafluoroborate ($[BF_4]^-$). Ultimately, 30 wt % [TBA][X] incorporation into PEIE layer at cathode interfaces improved WF as determined by ultraviolet photoelectron spectroscopy [232]. Researchers determined that anions with strong electron donating characteristics, such as $[SCN]^-$ and $[Ac]^-$, resulted in the largest reduction of WF, while small halides provided the lowest WF change.

Investigators continued their investigations with chain length studies using [TEA][Cl] and [THA][Cl] dopants in PEIE electrode coatings and studying WF values. Results indicated that longer chain lengths provided larger steric hindrance, and thus, weaker electron accepting ability. Overall, WF decreased as hypothesized; however, devices with PEIE:[TBA][SCN] doping resulted in an unexplained increase in drive voltage. In general, this work demonstrates anion influence on WF and electronic efficiencies in LEDs [232]. More recently, Duan and coworkers expanded this work to include anion exchange effects on polyelectrolytes inspired by PEIE design [233]. These investigators incorporated ammonium cations into the PEIE backbone and used several sulfonate anions, such as dimethyl sulfonate ($[MSB]^-$), benzylsulfonate ($[BSB]^-$), and diethyl sulfonate ($[ESB]^-$), to examine effects on WF for polymer solar cells. Notably, these researchers found that smaller anions, e.g. $[ESB]^-$ and $[MSB]^-$, yielded devices with more efficient electron transport characteristics and better performance than devices with $[BSB]^-$. Interestingly, devices with [PEIE][ESB] demonstrated the highest PCE (10.44%) with 8 nm thickness at minimal light soaking [233].

In another work, Sato and coworkers investigated counterion exchange effects on a polymerized IL system to reduce WF at the electron-injection layer to provide sufficient electrons to the semiconducting layer [231]. These investigators employed two fluorinated anions to produce polydiallylammonium polymeric ILs [poly(DDA)][NTf_2] and [poly(DDA)][BETI], respectively. Both polymers were evaluated and compared to their parent PIL [poly(DDA)][Cl]. Hydrophobicities were studied via water contact angle measurements between film samples and water droplets. As anticipated, larger contact angles were observed for the more hydrophobic anions $[NTf_2]^-$ and $[BETI]^-$ as compared with $[Cl]^-$, 80.1, 80.8 and 19.1°, respectively. Noticeable increases in 5% onset degradation temperatures of the polymers were also observed upon conversion from PIL to polymer-ion exchanges, from 285 °C in [poly(DDA)][Cl] to 394 and 395 °C with [poly(DDA)][NTf_2] and [poly(DDA)][BETI], respectively. Both ion-exchanged polymer-ILs reduced WF when they were incorporated at the cathode interface and increased WF when employed at the anode, which indicates that they are suitable interfacial coatings for OLED development. After device optimization as electron injection layers, the authors reported a best device performance of 9.00% maximum EQE with [poly(DDA)][NTf_2]. Overall, these researchers demonstrated the benefits of hydrophobic counterion exchange for PILs and their utility for OLEDs applications at electrode interfaces.

5. Conclusions and future directions

Much like room temperature ILs, the ionic properties of frozen ILs and GUMBOS lend to high tunability as a result of the exponential combinations possible of known anions and cations. This important characteristic allows for strategic design of specific GUMBOS for a targeted analytical task. This chapter summarized a few examples of possible GUMBOS applications. Moreover, it has been demonstrated that several physico-chemical properties of these compounds are improved in solid state as compared to liquid phase organic salts. For these reasons, we hypothesize that implementation of solid-phase ILs and GUMBOS in the analytical and materials fields will increase in the future.

Acknowledgements

The authors gratefully acknowledge financial support through NASA cooperative agreement NNX 16AQ93A under contract number NASA/LEQSF (2016-2019)-Phase 3-10, and the National Science Foundation under Grant Nos. CHE-1905105 and HRD-1736136. Any opinions, findings, and conclusions or recommendations expressed in this material are those of the author(s) and do not necessarily reflect the views of the National Science Foundation.

Conflict of interest

The authors declare no conflict of interest.

Author details

Rocío L. Pérez, Caitlan E. Ayala and Isiah M. Warner*
Department of Chemistry, Louisiana State University, Baton Rouge, LA, United States

*Address all correspondence to: iwarner@lsu.edu

References

[1] Brown, R.J. and M.J. Milton, *Analytical techniques for trace element analysis: an overview.* TrAC Trends in Analytical Chemistry, 2005. **24**(3): p. 266-274.

[2] Lin, D.C., E.K. Dimitriadis, and F. Horkay, *Robust strategies for automated AFM force curve analysis—I. Non-adhesive indentation of soft, inhomogeneous materials.* 2007.

[3] Dogan-Topal, B., S.A. Ozkan, and B. Uslu, *The analytical applications of square wave voltammetry on pharmaceutical analysis.* The Open Chemical and Biomedical Methods Journal, 2010. **3**(1).

[4] Chakraborty, P. and T. Pradeep, *The emerging interface of mass spectrometry with materials.* NPG Asia Materials, 2019. **11**(1): p. 1-22.

[5] Ambrose, W., et al., *A.; Werner, JH; Keller, RA.* Chem. ReV, 1999. **99**(10): p. 2929-2956.

[6] Pfannmoller, M., et al., *Visualizing a homogeneous blend in bulk heterojunction polymer solar cells by analytical electron microscopy.* Nano letters, 2011. **11**(8): p. 3099-3107.

[7] Commerce, N.I.o.S.a.T.-U.S.D.o., *Available online: Measurements, Standards, and Reference Materials for Industrial Commodities*. 2008.

[8] Ye, F., et al., *A compact and highly fluorescent orange-emitting polymer dot for specific subcellular imaging.* Chemical Communications, 2012. **48**(12): p. 1778-1780.

[9] Ramos, J., et al., *Soft nanoparticles (thermo-responsive nanogels and bicelles) with biotechnological applications: from synthesis to simulation through colloidal characterization.* Soft Matter, 2011. **7**(11): p. 5067-5082.

[10] Tessler, L.A., et al., *Nanogel surface coatings for improved single-molecule imaging substrates.* Journal of the Royal Society Interface, 2011. **8**(63): p. 1400-1408.

[11] Nishiyabu, R., et al., *Confining Molecules within Aqueous Coordination Nanoparticles by Adaptive Molecular Self-Assembly.* Angewandte Chemie, 2009. **121**(50): p. 9629-9632.

[12] Ojida, A., et al., *Turn-on fluorescence sensing of nucleoside polyphosphates using a xanthene-based Zn (II) complex chemosensor.* Journal of the American Chemical Society, 2008. **130**(36): p. 12095-12101.

[13] Cole, M.R.H., J.A.; Warner, I.M., *Recycling Antibiotics into GUMBOS: A New Combination Strategy to Combat Multi-Drug-Resistant Bacteria.* Molecules. **20**: p. 6466-6487.

[14] Cole, M.R., et al., *Minimizing human infection from Escherichia coli O157:H7 using GUMBOS.* Journal of Antimicrobial Chemotherapy, 2013. **68**(6): p. 1312-1318.

[15] Cole, M.R., et al., *Design, Synthesis, and Biological Evaluation of β-Lactam Antibiotic-Based Imidazolium- and Pyridinium-Type Ionic Liquids.* Chemical Biology & Drug Design, 2011. **78**(1): p. 33-41.

[16] Chen, M., et al., *Mitochondria targeting IR780-based nanoGUMBOS for enhanced selective toxicity towards cancer cells.* RSC Advances, 2018. **8**(55): p. 31700-31709.

[17] Bhattarai, N., et al., *Enhanced chemotherapeutic toxicity of cyclodextrin templated size-tunable rhodamine 6G nanoGUMBOS.* Journal of Materials Chemistry B, 2018. **6**(34): p. 5451-5459.

[18] Bhattarai, N., et al., *Endocytic Selective Toxicity of Rhodamine 6G*

nanoGUMBOS in Breast Cancer Cells. Molecular Pharmaceutics, 2018. **15**(9): p. 3837-3845.

[19] Das, S., et al., *Multimodal theranostic nanomaterials derived from phthalocyanine-based organic salt.* RSC Advances, 2015. **5**(38): p. 30227-30233.

[20] Siraj, N., et al., *Strategy for Tuning the Photophysical Properties of Photosensitizers for Use in Photodynamic Therapy.* Chemistry – A European Journal, 2015. **21**(41): p. 14440-14446.

[21] Dumke, J.C., et al., *In vitro activity studies of hyperthermal near-infrared nanoGUMBOS in MDA-MB-231 breast cancer cells.* Photochemical & Photobiological Sciences, 2014. **13**(9): p. 1270-1280.

[22] Magut, P.K.S., et al., *Tunable Cytotoxicity of Rhodamine 6G via Anion Variations.* Journal of the American Chemical Society, 2013. **135**(42): p. 15873-15879.

[23] Dumke, J.C., et al., *Photothermal Response of Near-Infrared-Absorbing NanoGUMBOS.* Applied Spectroscopy, 2014. **68**(3): p. 340-352.

[24] Li, M., et al., *Lipophilic phosphonium–lanthanide compounds with magnetic, luminescent, and tumor targeting properties.* Journal of Inorganic Biochemistry, 2012. **107**(1): p. 40-46.

[25] McNeel, K.E., et al., *Sodium Deoxycholate Hydrogels: Effects of Modifications on Gelation, Drug Release, and Nanotemplating.* The Journal of Physical Chemistry B, 2015. **119**(27): p. 8651-8659.

[26] McNeel, K.E., et al., *Sodium deoxycholate/TRIS-based hydrogels for multipurpose solute delivery vehicles: Ambient release, drug release, and enantiopreferential release.* Talanta, 2018. **177**: p. 66-73.

[27] McNeel, K.E., et al., *Fluorescence-Based Ratiometric Nanosensor for Selective Imaging of Cancer Cells.* ACS Omega, 2019. **4**(1): p. 1592-1600.

[28] Bwambok, D.K., et al., *Near-Infrared Fluorescent NanoGUMBOS for Biomedical Imaging.* ACS Nano, 2009. **3**(12): p. 3854-3860.

[29] De Rooy, S.L., et al., *Ephedrinium-based protic chiral ionic liquids for enantiomeric recognition.* Chirality, 2011. **23**(1): p. 54-62.

[30] Bwambok, D.K., et al., *Amino Acid-Based Fluorescent Chiral Ionic Liquid for Enantiomeric Recognition.* Analytical Chemistry, 2010. **82**(12): p. 5028-5037.

[31] Li, M., et al., *Combinatorial Approach to Enantiomeric Discrimination: Synthesis and 19F NMR Screening of a Chiral Ionic Liquid-Modified Silane Library.* Journal of Combinatorial Chemistry, 2009. **11**(6): p. 1105-1114.

[32] Li, M., et al., *Magnetic chiral ionic liquids derived from amino acids.* Chemical Communications, 2009(45): p. 6922-6924.

[33] Bwambok, D.K., et al., *Synthesis and characterization of novel chiral ionic liquids and investigation of their enantiomeric recognition properties.* Chirality, 2008. **20**(2): p. 151-158.

[34] Kolic, P.E., et al., *Improving energy relay dyes for dye-sensitized solar cells by use of a group of uniform materials based on organic salts (GUMBOS).* RSC Advances, 2016. **6**(97): p. 95273-95282.

[35] Kolic, P.E., et al., *Synthesis and Characterization of Porphyrin-Based GUMBOS and NanoGUMBOS as Improved Photosensitizers.* The Journal of Physical Chemistry C, 2016. **120**(9): p. 5155-5163.

[36] Jordan, A.N., et al., *Anion-controlled morphologies and spectral features of cyanine-based nanoGUMBOS – an improved photosensitizer.* Nanoscale, 2012. **4**(16): p. 5031-5038.

[37] Berton, P., et al., *Ionic liquid-based dispersive microextraction of nitrotoluenes in water samples.* Microchimica Acta, 2014. **181**(11): p. 1191-1198.

[38] Huang, F., et al., *Surfactant-based ionic liquids for extraction of phenolic compounds combined with rapid quantification using capillary electrophoresis.* ELECTROPHORESIS, 2014. **35**(17): p. 2463-2469.

[39] Deng, N., et al., *Highly efficient extraction of phenolic compounds by use of magnetic room temperature ionic liquids for environmental remediation.* Journal of Hazardous Materials, 2011. **192**(3): p. 1350-1357.

[40] Vidanapathirana, P., et al., *Cationic ionic liquid surfactant-polyacrylamide gel electrophoresis for enhanced separation of acidic and basic proteins with single-step ribonuclease b glycoforms separation.* Journal of Chromatography A, 2017. **1515**: p. 245-251.

[41] Hasan, F., et al., *Ionic liquids as buffer additives in ionic liquid-polyacrylamide gel electrophoresis separation of mixtures of low and high molecular weight proteins.* RSC Advances, 2015. **5**(85): p. 69229-69237.

[42] Cong, M., et al., *Ratiometric fluorescence detection of hydroxyl radical using cyanine-based binary nanoGUMBOS.* Sensors and Actuators B: Chemical, 2018. **257**: p. 993-1000.

[43] Lu, C., et al., *Spectral and Physicochemical Characterization of Dysprosium-Based Multifunctional Ionic Liquid Crystals.* The Journal of Physical Chemistry A, 2015. **119**(20): p. 4780-4786.

[44] Al Ghafly, H., et al., *GUMBOS matrices of variable hydrophobicity for matrix-assisted laser desorption/ionization mass spectrometry.* Rapid Communications in Mass Spectrometry, 2014. **28**(21): p. 2307-2314.

[45] Hamdan, S., et al., *Strategies for controlled synthesis of nanoparticles derived from a group of uniform materials based on organic salts.* Journal of Colloid and Interface Science, 2015. **446**: p. 163-169.

[46] Wright, A.R., et al., *Soft- and hard-templated organic salt nanoparticles with the Midas touch: gold-shelled nanoGUMBOS.* Journal of Materials Chemistry C, 2014. **2**(42): p. 8996-9003.

[47] de Rooy, S.L., et al., *Ionically Self-Assembled, Multi-Luminophore One-Dimensional Micro- and Nanoscale Aggregates of Thiacarbocyanine GUMBOS.* The Journal of Physical Chemistry C, 2012. **116**(14): p. 8251-8260.

[48] Das, S., et al., *Tunable Size and Spectral Properties of Fluorescent NanoGUMBOS in Modified Sodium Deoxycholate Hydrogels.* Langmuir, 2012. **28**(1): p. 757-765.

[49] de Rooy, S.L., et al., *Fluorescent one-dimensional nanostructures from a group of uniform materials based on organic salts.* Chemical Communications, 2011. **47**(31): p. 8916-8918.

[50] Das, S., et al., *Nontemplated Approach to Tuning the Spectral Properties of Cyanine-Based Fluorescent NanoGUMBOS.* Langmuir, 2010. **26**(15): p. 12867-12876.

[51] Tesfai, A., et al., *Controllable Formation of Ionic Liquid Micro- and Nanoparticles via a Melt–Emulsion–Quench Approach.* Nano Letters, 2008. **8**(3): p. 897-901.

[52] Tesfai, A., et al., *Magnetic and Nonmagnetic Nanoparticles from a Group of Uniform Materials Based on Organic Salts.* ACS Nano, 2009. **3**(10): p. 3244-3250.

[53] De Silva, T.P.D., et al., *Influence of Anion Variations on Morphological, Spectral, and Physical Properties of the Propidium Luminophore.* The Journal of Physical Chemistry A, 2019. **123**(1): p. 111-119.

[54] Siraj, N., et al., *Carbazole-Derived Group of Uniform Materials Based on Organic Salts: Solid State Fluorescent Analogues of Ionic Liquids for Potential Applications in Organic-Based Blue Light-Emitting Diodes.* The Journal of Physical Chemistry C, 2014. **118**(5): p. 2312-2320.

[55] Siraj, N., et al., *Enhanced S2 emission in carbazole-based ionic liquids.* RSC Advances, 2015. **5**(13): p. 9939-9945.

[56] Sarkar, A., et al., *Electro-optical characterization of cyanine-based GUMBOS and nanoGUMBOS.* Electronic Materials Letters, 2014. **10**(5): p. 879-885.

[57] Galpothdeniya, W.I.S., et al., *Fluorescein-based ionic liquid sensor for label-free detection of serum albumins.* RSC Advances, 2014. **4**(34): p. 17533-17540.

[58] Galpothdeniya, W.I.S., et al., *Ionic liquid-based optoelectronic sensor arrays for chemical detection.* RSC Advances, 2014. **4**(14): p. 7225-7234.

[59] Galpothdeniya, W.I.S., et al., *Tunable GUMBOS-based sensor array for label-free detection and discrimination of proteins.* Journal of Materials Chemistry B, 2016. **4**(8): p. 1414-1422.

[60] Das, S., et al., *Ionic liquid-based fluorescein colorimetric pH nanosensors.* RSC Advances, 2013. **3**(43): p. 21054-21061.

[61] Galpothdeniya, W.I.S., et al., *Virtual Colorimetric Sensor Array: Single Ionic Liquid for Solvent Discrimination.* Analytical Chemistry, 2015. **87**(8): p. 4464-4471.

[62] Stavrou, I.J., et al., *Facile preparation of polysaccharide-coated capillaries using a room temperature ionic liquid for chiral separations.* ELECTROPHORESIS, 2013. **34**(9-10): p. 1334-1338.

[63] Hamdan, S., et al., *Ionic liquid crosslinkers for chiral imprinted nanoGUMBOS.* Journal of Colloid and Interface Science, 2016. **463**: p. 29-36.

[64] Karam, T.E., et al., *Anomalous Size-Dependent Excited-State Relaxation Dynamics of NanoGUMBOS.* The Journal of Physical Chemistry C, 2015. **119**(50): p. 28206-28213.

[65] Karam, T.E., et al., *Ultrafast and nonlinear spectroscopy of brilliant green-based nanoGUMBOS with enhanced near-infrared emission.* The Journal of Chemical Physics, 2017. **147**(14): p. 144701.

[66] Jordan, A.N., et al., *Tunable near-infrared emission of binary nano- and mesoscale GUMBOS.* RSC Advances, 2014. **4**(54): p. 28471-28480.

[67] Lu, C., et al., *Irradiation Induced Fluorescence Enhancement in PEGylated Cyanine-Based NIR Nano- and Mesoscale GUMBOS.* Langmuir, 2012. **28**(40): p. 14415-14423.

[68] Vaughan, S.R., et al., *Class specific discrimination of volatile organic compounds using a quartz crystal microbalance based multisensor array.* Talanta, 2018. **188**: p. 423-428.

[69] Speller, N.C., et al., *QCM virtual sensor array: Vapor identification and molecular weight approximation.* Sensors and Actuators B: Chemical, 2017. **246**: p. 952-960.

[70] Speller, N.C., et al., *QCM virtual multisensor array for fuel discrimination and detection of gasoline adulteration.* Fuel, 2017. **199**: p. 38-46.

[71] Speller, N.C., et al., *Assessment of QCM array schemes for mixture identification: citrus scented odors.* RSC Advances, 2016. **6**(98): p. 95378-95386.

[72] Speller, N.C., et al., *Rational Design of QCM-D Virtual Sensor Arrays Based on Film Thickness, Viscoelasticity, and Harmonics for Vapor Discrimination.* Analytical Chemistry, 2015. **87**(10): p. 5156-5166.

[73] Regmi, B.P., et al., *Phthalocyanine- and porphyrin-based GUMBOS for rapid and sensitive detection of organic vapors.* Sensors and Actuators B: Chemical, 2015. **209**: p. 172-179.

[74] Regmi, B.P., et al., *Molecular weight sensing properties of ionic liquid-polymer composite films: theory and experiment.* Journal of Materials Chemistry C, 2014. **2**(24): p. 4867-4878.

[75] Regmi, B.P., et al., *A novel composite film for detection and molecular weight determination of organic vapors.* Journal of Materials Chemistry, 2012. **22**(27): p. 13732-13741.

[76] Siraj, N.W., Isiah M. , *Compositions including a ruthenium molecular dye-based GUMBOS, methods of making compositions, methods of use of compositions, and devices using the compositions*. 2015.

[77] Siraj, N.W., Isiah M., *Carbazole based GUMBOS for potential applications as highly efficient Blue OLED's*.

[78] Warner, I.M., Regmi, Bishnu P., El-Zahab, Bilal, Hayes, Daniel J. , *Detection and Molecular Weight Estimation of Organic Vapors Using a QCM Sensor*. 2012.

[79] Trujillo-Rodríguez, M.J., et al., *Advances of Ionic Liquids in Analytical Chemistry.* Analytical Chemistry, 2019. **91**(1): p. 505-531.

[80] Liu, Q., S.Z. El Abedin, and F. Endres, *Electroplating of mild steel by aluminium in a first generation ionic liquid: A green alternative to commercial Al-plating in organic solvents.* Surface and Coatings Technology, 2006. **201**(3-4): p. 1352-1356.

[81] Bai, L., et al., *Effects of nucleators on the thermodynamic properties of seasonal energy storage materials based on ionic liquids.* Energy & Fuels, 2011. **25**(4): p. 1811-1816.

[82] Vekariya, R.L., *A review of ionic liquids: Applications towards catalytic organic transformations.* Journal of Molecular Liquids, 2017. **227**: p. 44-60.

[83] Qiao, Y., et al., *Temperature-responsive ionic liquids: fundamental behaviors and catalytic applications.* Chemical reviews, 2017. **117**(10): p. 6881-6928.

[84] Stracke, M.P., et al., *Hydrogen-storage materials based on imidazolium ionic liquids.* Energy & fuels, 2007. **21**(3): p. 1695-1698.

[85] Mahrova, M., et al., *Pyridinium based dicationic ionic liquids as base lubricants or lubricant additives.* Tribology International, 2015. **82**: p. 245-254.

[86] Zhou, Y. and J. Qu, *Ionic liquids as lubricant additives: a review.* ACS applied materials & interfaces, 2017. **9**(4): p. 3209-3222.

[87] Egorova, K.S., E.G. Gordeev, and V.P. Ananikov, *Biological activity of ionic liquids and their application in pharmaceutics and medicine.* Chemical Reviews, 2017. **117**(10): p. 7132-7189.

[88] Gomes, J.M., S.S. Silva, and R.L. Reis, *Biocompatible ionic liquids: fundamental behaviours and applications.*

Chemical Society Reviews, 2019. **48**(15): p. 4317-4335.

[89] Chen, M., et al., *Tumor-Targeting NIRF NanoGUMBOS with Cyclodextrin-Enhanced Chemo/Photothermal Antitumor Activities.* ACS applied materials & interfaces, 2019. **11**(31): p. 27548-27557.

[90] Warner, I.M., B. El-Zahab, and N. Siraj, *Perspectives on Moving Ionic Liquid Chemistry into the Solid Phase.* Analytical Chemistry, 2014. **86**(15): p. 7184-7191.

[91] MacFarlane, D.R., et al., *Ionic liquids and their solid-state analogues as materials for energy generation and storage.* Nature Reviews Materials, 2016. **1**(2): p. 1-15.

[92] Siegel, R.L., K.D. Miller, and A. Jemal, *Cancer statistics, 2020.* CA: A Cancer Journal for Clinicians, 2020. **70**(1): p. 7-30.

[93] Miller, K.D., et al., *Cancer treatment and survivorship statistics, 2019.* CA: a cancer journal for clinicians, 2019. **69**(5): p. 363-385.

[94] Pearce, A., et al., *Incidence and severity of self-reported chemotherapy side effects in routine care: A prospective cohort study.* PloS one, 2017. **12**(10): p. e0184360.

[95] Nurgali, K., R.T. Jagoe, and R. Abalo, *Adverse effects of cancer chemotherapy: Anything new to improve tolerance and reduce sequelae?* Frontiers in pharmacology, 2018. **9**: p. 245.

[96] Oun, R., Y.E. Moussa, and N.J. Wheate, *The side effects of platinum-based chemotherapy drugs: a review for chemists.* Dalton Transactions, 2018. **47**(19): p. 6645-6653.

[97] Zhang, J., et al., *An updated overview on the development of new photosensitizers for anticancer photodynamic therapy.* Acta Pharmaceutica Sinica B, 2018. **8**(2): p. 137-146.

[98] Sharma, R.A., et al., *Clinical development of new drug–radiotherapy combinations.* Nature reviews Clinical oncology, 2016. **13**(10): p. 627-642.

[99] Pérez-Herrero, E. and A. Fernández-Medarde, *Advanced targeted therapies in cancer: Drug nanocarriers, the future of chemotherapy.* European Journal of Pharmaceutics and Biopharmaceutics, 2015. **93**: p. 52-79.

[100] Li, W., et al., *Mild photothermal therapy/photodynamic therapy/ chemotherapy of breast cancer by Lyp-1 modified Docetaxel/IR820 Co-loaded micelles.* Biomaterials, 2016. **106**: p. 119-133.

[101] Kim, H.S. and D.Y. Lee, *Near-infrared-responsive cancer photothermal and photodynamic therapy using gold nanoparticles.* Polymers, 2018. **10**(9): p. 961.

[102] Paris, J.L., et al., *Nanoparticles for multimodal antivascular therapeutics: Dual drug release, photothermal and photodynamic therapy.* Acta Biomaterialia, 2020. **101**: p. 459-468.

[103] Xu, F., et al., *Loading of indocyanine green within polydopamine-coated laponite nanodisks for targeted cancer photothermal and photodynamic therapy.* Nanomaterials, 2018. **8**(5): p. 347.

[104] Pedersen, P.L., *Tumor mitochondria and the bioenergetics of cancer cells*, in *Membrane anomalies of tumor cells*. 1978, Karger Publishers. p. 190-274.

[105] Carracedo, A., L.C. Cantley, and P.P. Pandolfi, *Cancer metabolism: fatty acid oxidation in the limelight.* Nature reviews Cancer, 2013. **13**(4): p. 227-232.

[106] Liberti, M.V. and J.W. Locasale, *The Warburg Effect: How Does it Benefit Cancer Cells?* Trends in Biochemical Sciences, 2016. **41**(3): p. 211-218.

[107] Nödling, A.R., et al., *Cyanine dye mediated mitochondrial targeting enhances the anti-cancer activity of small-molecule cargoes.* Chemical Communications, 2020. **56**(34): p. 4672-4675.

[108] Bhattarai, N., et al., *Endocytic selective toxicity of rhodamine 6G nanoGUMBOS in breast cancer cells.* Molecular pharmaceutics, 2018. **15**(9): p. 3837-3845.

[109] Gear, A.R., *Rhodamine 6G a potent inhibitor of mitochondrial oxidative phosphorylation.* Journal of Biological Chemistry, 1974. **249**(11): p. 3628-3637.

[110] Modica-Napolitano, J.S., et al., *Rhodamine 123 inhibits bioenergetic function in isolated rat liver mitochondria.* Biochemical and biophysical research communications, 1984. **118**(3): p. 717-723.

[111] Davis, S., et al., *Mitochondrial and plasma membrane potentials cause unusual accumulation and retention of rhodamine 123 by human breast adenocarcinoma-derived MCF-7 cells.* Journal of Biological Chemistry, 1985. **260**(25): p. 13844-13850.

[112] Modica-Napolitano, J.S. and J.R. Aprille, *Basis for the selective cytotoxicity of rhodamine 123.* Cancer research, 1987. **47**(16): p. 4361-4365.

[113] Scaduto Jr, R.C. and L.W. Grotyohann, *Measurement of mitochondrial membrane potential using fluorescent rhodamine derivatives.* Biophysical journal, 1999. **76**(1): p. 469-477.

[114] Ma, X., et al., *A Mitochondria-Targeting Gold–Peptide Nanoassembly for Enhanced Cancer-Cell Killing.* Advanced healthcare materials, 2013. **2**(12): p. 1638-1643.

[115] Hu, Y.-P., et al., *ρ0 tumor cells: a model for studying whether mitochondria are targets for rhodamine 123, doxorubicin, and other drugs.* Biochemical Pharmacology, 2000. **60**(12): p. 1897-1905.

[116] Alford, R., et al., *Toxicity of organic fluorophores used in molecular imaging: literature review.* Molecular imaging, 2009. **8**(6): p. 7290.2009. 00031.

[117] Mosquera, J.s., I. García, and L.M. Liz-Marzán, *Cellular uptake of nanoparticles versus small molecules: a matter of size.* Accounts of chemical research, 2018. **51**(9): p. 2305-2313.

[118] Ernsting, M.J., et al., *Factors controlling the pharmacokinetics, biodistribution and intratumoral penetration of nanoparticles.* Journal of Controlled Release, 2013. **172**(3): p. 782-794.

[119] Duan, X. and Y. Li, *Physicochemical characteristics of nanoparticles affect circulation, biodistribution, cellular internalization, and trafficking.* Small, 2013. **9**(9-10): p. 1521-1532.

[120] Stylianopoulos, T., *EPR-effect: utilizing size-dependent nanoparticle delivery to solid tumors.* Therapeutic delivery, 2013. **4**(4): p. 421-423.

[121] Loftsson, T., et al., *Cyclodextrins in drug delivery.* Expert opinion on drug delivery, 2005. **2**(2): p. 335-351.

[122] Miranda, J.C.d., et al., *Cyclodextrins and ternary complexes: technology to improve solubility of poorly soluble drugs.* Brazilian journal of pharmaceutical sciences, 2011. **47**(4): p. 665-681.

[123] Kuang, Y., et al., *Hydrophobic IR-780 dye encapsulated in cRGD-conjugated solid lipid nanoparticles for NIR imaging-guided photothermal therapy.* ACS Applied Materials & Interfaces, 2017. **9**(14): p. 12217-12226.

[124] Chen, Y., et al., *IR-780 loaded phospholipid mimicking homopolymeric*

micelles for near-IR imaging and photothermal therapy of pancreatic cancer. ACS applied materials & interfaces, 2016. **8**(11): p. 6852-6858.

[125] Xia, F., et al., *Matrix metallopeptidase 2 targeted delivery of gold nanostars decorated with IR-780 iodide for dual-modal imaging and enhanced photothermal/photodynamic therapy.* Acta biomaterialia, 2019. **89**: p. 289-299.

[126] Zhang, C., et al., *A near-infrared fluorescent heptamethine indocyanine dye with preferential tumor accumulation for in vivo imaging.* Biomaterials, 2010. **31**(25): p. 6612-6617.

[127] Choi, J., et al., *Targeting tumors with cyclic RGD-conjugated lipid nanoparticles loaded with an IR780 NIR dye: In vitro and in vivo evaluation.* International Journal of Pharmaceutics, 2017. **532**(2): p. 677-685.

[128] Dai, Q., C. Walkey, and W.C. Chan, *Polyethylene glycol backfilling mitigates the negative impact of the protein corona on nanoparticle cell targeting.* Angewandte Chemie International Edition, 2014. **53**(20): p. 5093-5096.

[129] Palao-Suay, R., et al., *Photothermal and photodynamic activity of polymeric nanoparticles based on α-tocopheryl succinate-RAFT block copolymers conjugated to IR-780.* Acta biomaterialia, 2017. **57**: p. 70-84.

[130] Gu, F.X., et al., *Targeted nanoparticles for cancer therapy.* Nano Today, 2007. **2**(3): p. 14-21.

[131] Broadwater, D., et al., *Modulating cellular cytotoxicity and phototoxicity of fluorescent organic salts through counterion pairing.* Scientific Reports, 2019. **9**(1): p. 15288.

[132] Sneader, W., *History of Sulfonamides.* e LS, 2001.

[133] Gaynes, R., *The Discovery of Penicillin—New Insights After More Than 75 Years of Clinical Use.* Emerging Infectious Diseases, 2017. **23**(5): p. 849-853.

[134] Woodruff, H.B., *Selman A. Waksman, winner of the 1952 Nobel Prize for physiology or medicine.* Applied and environmental microbiology, 2014. **80**(1): p. 2-8.

[135] Davies, J., *Microbes have the last word: A drastic re-evaluation of antimicrobial treatment is needed to overcome the threat of antibiotic-resistant bacteria.* EMBO reports, 2007. **8**(7): p. 616-621.

[136] Durand, G.A., D. Raoult, and G. Dubourg, *Antibiotic discovery: history, methods and perspectives.* International Journal of Antimicrobial Agents, 2019. **53**(4): p. 371-382.

[137] Llor, C. and L. Bjerrum, *Antimicrobial resistance: risk associated with antibiotic overuse and initiatives to reduce the problem.* Therapeutic advances in drug safety, 2014. **5**(6): p. 229-241.

[138] Shallcross, L.J. and D.S.C. Davies, *Antibiotic overuse: a key driver of antimicrobial resistance.* 2014, British Journal of General Practice.

[139] Zaman, S.B., et al., *A Review on Antibiotic Resistance: Alarm Bells are Ringing.* Cureus, 2017. **9**(6): p. e1403-e1403.

[140] Liu, F. and A.G. Myers, *Development of a platform for the discovery and practical synthesis of new tetracycline antibiotics.* Current Opinion in Chemical Biology, 2016. **32**: p. 48-57.

[141] Bekri, S., et al., *New antibacterial cadiolide analogues active against antibiotic-resistant strains.* Bioorganic & Medicinal Chemistry Letters, 2020. **30**(21): p. 127580.

[142] Dorst, A., et al., *Semisynthetic Analogs of the Antibiotic Fidaxomicin—Design, Synthesis, and Biological Evaluation.* ACS Medicinal Chemistry Letters, 2020.

[143] Florindo, C., et al., *Evaluation of solubility and partition properties of ampicillin-based ionic liquids.* International Journal of Pharmaceutics, 2013. **456**(2): p. 553-559.

[144] Florindo, C., et al., *Novel organic salts based on fluoroquinolone drugs: Synthesis, bioavailability and toxicological profiles.* International Journal of Pharmaceutics, 2014. **469**(1): p. 179-189.

[145] Santos, M.M., et al., *Antimicrobial Activities of Highly Bioavailable Organic Salts and Ionic Liquids from Fluoroquinolones.* Pharmaceutics, 2020. **12**(8): p. 694.

[146] Frizzo, C.P., et al., *Novel ibuprofenate- and docusate-based ionic liquids: emergence of antimicrobial activity.* RSC Advances, 2016. **6**(102): p. 100476-100486.

[147] Ferraz, R., et al., *Synthesis and Antibacterial Activity of Ionic Liquids and Organic Salts based on Penicillin G and Amoxicillin hydrolysate derivatives against Resistant Bacteria.* Pharmaceutics, 2020. **12**(3): p. 221.

[148] Rangel, J.M., et al., *Epidemiology of Escherichia coli O157: H7 outbreaks, united states, 1982-2002.* Emerging infectious diseases, 2005. **11**(4): p. 603.

[149] Gobin, M., et al., *National outbreak of Shiga toxin-producing Escherichia coli O157: H7 linked to mixed salad leaves, United Kingdom, 2016.* Eurosurveillance, 2018. **23**(18): p. 17-00197.

[150] Puño-Sarmiento, J., et al., *Potentiation of Antibiotics by a Novel Antimicrobial Peptide against Shiga Toxin Producing E. coli O157: H7.* Scientific reports, 2020. **10**(1): p. 1-14.

[151] Cole, M.R., J.A. Hobden, and I.M. Warner, *Recycling antibiotics into GUMBOS: a new combination strategy to combat multi-drug-resistant bacteria.* Molecules, 2015. **20**(4): p. 6466-6487.

[152] Cole, M., J. Hobden, and I. Warner, *Recycling Antibiotics into GUMBOS: A New Combination Strategy to Combat Multi-Drug-Resistant Bacteria.* Molecules, 2015. **20**(4): p. 6466.

[153] Newman, L.M., J.S. Moran, and K.A. Workowski, *Update on the Management of Gonorrhea in Adults in the United States.* Clinical Infectious Diseases, 2007. **44**(Supplement_3): p. S84-S101.

[154] Wi, T., et al., *Antimicrobial resistance in Neisseria gonorrhoeae: global surveillance and a call for international collaborative action.* PLoS medicine, 2017. **14**(7): p. e1002344.

[155] Lahra, M.M., et al., *Cooperative recognition of internationally disseminated ceftriaxone-resistant Neisseria gonorrhoeae strain.* Emerging infectious diseases, 2018. **24**(4): p. 735.

[156] Lopez, K.M., J.A. Hobden, and I.M. Warner, *Octenidine/carbenicillin GUMBOS as potential treatment for oropharyngeal gonorrhoea.* Journal of Antimicrobial Chemotherapy, 2020.

[157] Wu, S., et al., *Counterions-mediated gold nanorods-based sensor for label-free detection of poly(ADP-ribose) polymerase-1 activity and its inhibitor.* Sensors and Actuators B: Chemical, 2018. **259**: p. 565-572.

[158] Li, H.-Y.C., Y.-H., *Reaction-Based Amine and Alcohol Gases Detection with Triazine Ionic Liquid Materials.* Molecules, 2020. **25**: p. 104.

[159] Hsu, T.-H., S.-J. Chiang, and Y.-H. Chu, *Quartz Crystal Microbalance Analysis of Diels–Alder Reactions of Alkene Gases to Functional Ionic Liquids*

on Chips. Analytical Chemistry, 2016. **88**(22): p. 10837-10841.

[160] Li, G., et al., *Cation–Anion Interaction-Directed Molecular Design Strategy for Mechanochromic Luminescence.* Advanced Functional Materials, 2014. **24**(6): p. 747-753.

[161] Taynton, P., et al., *Repairable woven carbon fiber composites with full recyclability enabled by malleable polyimine networks.* Advanced Materials, 2016. **28**(15): p. 2904-2909.

[162] Li, G., et al., *Cation–anion interaction directed dual-mode switchable mechanochromic luminescence.* Journal of Materials Chemistry C, 2017. **5**(33): p. 8527-8534.

[163] Guimarães, L.B., et al., *Highly sensitive and precise optical temperature sensors based on new luminescent Tb3+/Eu3+ tetrakis complexes with imidazolic counterions.* Materials Advances, 2020. **1**(6): p. 1988-1995.

[164] You, L., D. Zha, and E.V. Anslyn, *Recent Advances in Supramolecular Analytical Chemistry Using Optical Sensing.* Chemical Reviews, 2015. **115**(15): p. 7840-7892.

[165] Chen, C.-Y., K.-H. Li, and Y.-H. Chu, *Reaction-Based Detection of Chemical Warfare Agent Mimics with Affinity Ionic Liquids.* Analytical Chemistry, 2018. **90**(14): p. 8320-8325.

[166] Zhang, W., et al., *AIE-doped poly(ionic liquid) photonic spheres: a single sphere-based customizable sensing platform for the discrimination of multi-analytes.* Chemical Science, 2017. **8**(9): p. 6281-6289.

[167] Achyuthan, K.E., et al., *Fluorescence superquenching of conjugated polyelectrolytes: applications for biosensing and drug discovery.* Journal of Materials Chemistry, 2005. **15**(27-28): p. 2648-2656.

[168] Thomas, S.W., G.D. Joly, and T.M. Swager, *Chemical Sensors Based on Amplifying Fluorescent Conjugated Polymers.* Chemical Reviews, 2007. **107**(4): p. 1339-1386.

[169] Jones, R.M., et al., *Superquenching and Its Applications in J-Aggregated Cyanine Polymers.* Langmuir, 2001. **17**(9): p. 2568-2571.

[170] Dickson, R.M., et al., *On/off blinking and switching behaviour of single molecules of green fluorescent protein.* Nature, 1997. **388**(6640): p. 355-358.

[171] Lin, H., et al., *Collective Fluorescence Blinking in Linear J-Aggregates Assisted by Long-Distance Exciton Migration.* Nano Letters, 2010. **10**(2): p. 620-626.

[172] Reisch, A., et al., *Collective fluorescence switching of counterion-assembled dyes in polymer nanoparticles.* Nature Communications, 2014. **5**(1): p. 4089.

[173] Würthner, F., *Perylene bisimide dyes as versatile building blocks for functional supramolecular architectures.* Chemical Communications, 2004(14): p. 1564-1579.

[174] Klymchenko, A.S., et al., *Highly lipophilic fluorescent dyes in nano-emulsions: towards bright non-leaking nano-droplets.* RSC Advances, 2012. **2**(31): p. 11876-11886.

[175] Shulov, I., et al., *Fluorinated counterion-enhanced emission of rhodamine aggregates: ultrabright nanoparticles for bioimaging and light-harvesting.* Nanoscale, 2015. **7**(43): p. 18198-18210.

[176] Andreiuk, B., et al., *An aluminium-based fluorinated counterion for enhanced encapsulation and emission of dyes in biodegradable polymer nanoparticles.* Materials Chemistry Frontiers, 2017. **1**(11): p. 2309-2316.

[177] Severi, C., N. Melnychuk, and A.S. Klymchenko, *Smartphone-assisted detection of nucleic acids by light-harvesting FRET-based nanoprobe.* Biosensors and Bioelectronics, 2020. **168**: p. 112515.

[178] Hong, S.W. and W.H. Jo, *A fluorescence resonance energy transfer probe for sensing pH in aqueous solution.* Polymer, 2008. **49**(19): p. 4180-4187.

[179] Ashokkumar, P., N. Adarsh, and A.S. Klymchenko, *Ratiometric Nanoparticle Probe Based on FRET-Amplified Phosphorescence for Oxygen Sensing with Minimal Phototoxicity.* Small, 2020. **16**(32): p. 2002494.

[180] Stoeva, S.I., et al., *Multiplexed Detection of Protein Cancer Markers with Biobarcoded Nanoparticle Probes.* Journal of the American Chemical Society, 2006. **128**(26): p. 8378-8379.

[181] Kumar, V., et al., *Nanostructured Aptamer-Functionalized Black Phosphorus Sensing Platform for Label-Free Detection of Myoglobin, a Cardiovascular Disease Biomarker.* ACS Applied Materials & Interfaces, 2016. **8**(35): p. 22860-22868.

[182] Pérez, R.L., et al., *Protein Discrimination Using a Fluorescence-Based Sensor Array of Thiacarbocyanine-GUMBOS.* ACS Sensors, 2020. **5**(8): p. 2422-2429.

[183] Hornbeck, P., *Enzyme-Linked Immunosorbent Assays.* Current Protocols in Immunology, 1992. **1**(1): p. 2.1.1-2.1.22.

[184] Ambrosi, A., F. Airò, and A. Merkoçi, *Enhanced Gold Nanoparticle Based ELISA for a Breast Cancer Biomarker.* Analytical Chemistry, 2010. **82**(3): p. 1151-1156.

[185] Pandey, S., *Analytical applications of room-temperature ionic liquids: A review of recent efforts.* Analytica Chimica Acta, 2006. **556**(1): p. 38-45.

[186] Goubaidoulline, I., G. Vidrich, and D. Johannsmann, *Organic Vapor Sensing with Ionic Liquids Entrapped in Alumina Nanopores on Quartz Crystal Resonators.* Analytical Chemistry, 2005. 77(2): p. 615-619.

[187] Gebicki, J., *Application of ionic liquids in electronic nose instruments*, in *Analytical Applications of Ionic Liquids*. 2016, WORLD SCIENTIFIC (EUROPE). p. 339-360.

[188] Park, C.H., et al., *Ionic Liquid-Carbon Nanotube Sensor Arrays for Human Breath Related Volatile Organic Compounds.* ACS Sensors, 2018. **3**(11): p. 2432-2437.

[189] Vaughan, S.R.P., R.L.; Chhotaray, P.; Warner, I.M., *Quartz Crystal Microbalance Based Sensor Arrays for Detection and Discrimination of VOCs Using Phosphonium Ionic Liquid Composites.* Sensors and Actuators B: Chemical, 2020. **20**: p. 615.

[190] Eritt, M., et al., *OLED manufacturing for large area lighting applications.* Thin Solid Films, 2010. **518**(11): p. 3042-3045.

[191] Evans, O.R. and W. Lin, *Crystal Engineering of NLO Materials Based on Metal–Organic Coordination Networks.* Accounts of Chemical Research, 2002. **35**(7): p. 511-522.

[192] Moorthy, J.N., et al., *De Novo Design for Functional Amorphous Materials: Synthesis and Thermal and Light-Emitting Properties of Twisted Anthracene-Functionalized Bimesitylenes.* Journal of the American Chemical Society, 2008. **130**(51): p. 17320-17333.

[193] Jeong, E.G., et al., *A review of highly reliable flexible encapsulation technologies towards rollable and foldable OLEDs.* Journal of Information Display, 2020. **21**(1): p. 19-32.

[194] Justin Thomas, K.R., et al., *Light-Emitting Carbazole Derivatives: Potential Electroluminescent Materials.* Journal of the American Chemical Society, 2001. **123**(38): p. 9404-9411.

[195] Ma, D., Y. Qiu, and L. Duan, *New Insights into Tunable Volatility of Ionic Materials through Counter-Ion Control.* Advanced Functional Materials, 2016. **26**(20): p. 3438-3445.

[196] Baldo, M.A., et al., *Highly efficient phosphorescent emission from organic electroluminescent devices.* Nature, 1998. **395**(6698): p. 151-154.

[197] Thejo Kalyani, N. and S.J. Dhoble, *Organic light emitting diodes: Energy saving lighting technology—A review.* Renewable and Sustainable Energy Reviews, 2012. **16**(5): p. 2696-2723.

[198] Ma, D., L. Duan, and Y. Qiu, *Orange-red- and white-emitting diodes fabricated by vacuum evaporation deposition of sublimable cationic iridium complexes.* Journal of Materials Chemistry C, 2016. **4**(22): p. 5051-5058.

[199] Bai, R., et al., *Sky-blue-emitting cationic iridium complexes with oxadiazole/triazine-type counter-anions and their use for efficient solution-processed organic light-emitting diodes.* Dyes and Pigments, 2021. **184**: p. 108586.

[200] Tang, F., et al., *A sky-blue fluorescent small molecule for non-doped OLED using solution-processing.* RSC Advances, 2015. **5**(87): p. 71419-71424.

[201] Merklein, L.D., D.; Braig, F.; Schlisske, S.; Rödlmeier, T.; Mink, M.; Kourkoulos, D.; Ulber, B.; Di Biase, M.; Meerholz, K.; Hernandez-Sosa, G.; Lemmer, U.; Sauer, H.M.; Dörsam, E.; Scharfer, P.; Schabel, W., *Comparative Study of Printed Multilayer OLED Fabrication through Slot Die Coating, Gravure and Inkjet Printing, and Their Combination.* Colloids Interfaces, 2019. **3**: p. 32.

[202] Zhou, L., et al., *Inkjet-Printed Small-Molecule Organic Light-Emitting Diodes: Halogen-Free Inks, Printing Optimization, and Large-Area Patterning.* ACS Applied Materials & Interfaces, 2017. **9**(46): p. 40533-40540.

[203] Guo, F., et al., *The fabrication of color-tunable organic light-emitting diode displays via solution processing.* Light: Science & Applications, 2017. **6**(11): p. e17094-e17094.

[204] Chiba, T., Y.-J. Pu, and J. Kido, *Solution-processable electron injection materials for organic light-emitting devices.* Journal of Materials Chemistry C, 2015. **3**(44): p. 11567-11576.

[205] Zhang, G., et al., *Anion-controlled dimer distance induced unique solid-state fluorescence of cyano substituted styrene pyridinium.* Scientific Reports, 2016. **6**(1): p. 37609.

[206] Cosa, G., et al., *Photophysical Properties of Fluorescent DNA-dyes Bound to Single- and Double-stranded DNA in Aqueous Buffered Solution¶.* Photochemistry and Photobiology, 2001. **73**(6): p. 585-599.

[207] De Silva, T.P.D., et al., *Pyrenylpyridines: Sky-Blue Emitters for Organic Light-Emitting Diodes.* ACS Omega, 2019. **4**(16): p. 16867-16877.

[208] Cole, J.M., et al., *Cosensitization in Dye-Sensitized Solar Cells.* Chemical Reviews, 2019. **119**(12): p. 7279-7327.

[209] Pagliaro, M., R. Ciriminna, and G. Palmisano, *BIPV: merging the photovoltaic with the construction industry.* Progress in Photovoltaics: Research and Applications, 2010. **18**(1): p. 61-72.

[210] Zhang, K., et al., *High-Performance, Transparent, Dye-Sensitized Solar Cells for See-Through Photovoltaic Windows.* Advanced Energy Materials, 2014. **4**(11): p. 1301966.

[211] Mohamad, A.A., *Physical properties of quasi-solid-state polymer electrolytes for dye-sensitised solar cells: A characterisation review.* Solar Energy, 2019. **190**: p. 434-452.

[212] Kuang, D., et al., *High Molar Extinction Coefficient Ion-Coordinating Ruthenium Sensitizer for Efficient and Stable Mesoscopic Dye-Sensitized Solar Cells.* Advanced Functional Materials, 2007. **17**(1): p. 154-160.

[213] Wang, P., et al., *A stable quasi-solid-state dye-sensitized solar cell with an amphiphilic ruthenium sensitizer and polymer gel electrolyte.* Nature Materials, 2003. **2**(6): p. 402-407.

[214] Abbotto, A., et al., *Panchromatic ruthenium sensitizer based on electron-rich heteroarylvinylene π-conjugated quaterpyridine for dye-sensitized solar cells.* Dalton Transactions, 2011. **40**(1): p. 234-242.

[215] Jradi, F.M., et al., *Near-Infrared Asymmetrical Squaraine Sensitizers for Highly Efficient Dye Sensitized Solar Cells: The Effect of π-Bridges and Anchoring Groups on Solar Cell Performance.* Chemistry of Materials, 2015. **27**(7): p. 2480-2487.

[216] Saccone, D., et al., *Polymethine Dyes in Hybrid Photovoltaics: Structure–Properties Relationships.* European Journal of Organic Chemistry, 2016. **2016**(13): p. 2244-2259.

[217] Osedach, T.P., et al., *Near-infrared photodetector consisting of J-aggregating cyanine dye and metal oxide thin films.* Applied Physics Letters, 2012. **101**(11): p. 113303.

[218] Hales, J.M., et al., *Design of Polymethine Dyes with Large Third-Order Optical Nonlinearities and Loss Figures of Merit.* Science, 2010. **327**(5972): p. 1485.

[219] Zhao, Y., et al., *Near-Infrared Harvesting Transparent Luminescent Solar Concentrators.* Advanced Optical Materials, 2014. **2**(7): p. 606-611.

[220] Fan, B., et al., *High performing doped cyanine bilayer solar cell.* Organic Electronics, 2010. **11**(4): p. 583-588.

[221] Eskandari, M., et al., *Counterion-Mediated Crossing of the Cyanine Limit in Crystals and Fluid Solution: Bond Length Alternation and Spectral Broadening Unveiled by Quantum Chemistry.* Journal of the American Chemical Society, 2020. **142**(6): p. 2835-2843.

[222] Gayton, J.N.A., S.; Fortenberry, R.C.; Hammer, N.I.; Delcamp, J.H., *Counter Anion Effect on the Photophysical Properties of Emissive Indolizine-Cyanine Dyes in Solution and Solid State.* Molecules, 2018. **23**: p. 3051.

[223] Huckaba, A.J., et al., *A low recombination rate indolizine sensitizer for dye-sensitized solar cells.* Chemical Communications, 2016. **52**(54): p. 8424-8427.

[224] Glinka, A., et al., *Interface Modification and Exceptionally Fast Regeneration in Copper Mediated Solar Cells Sensitized with Indoline Dyes.* The Journal of Physical Chemistry C, 2020. **124**(5): p. 2895-2906.

[225] Abu Talip, R.A.Y., W.Z.N.; Bustam, M.A, *Ionic Liquids Roles and Perspectives in Electrolyte for Dye-Sensitized Solar Cells.* Sustainability, 2020. **12**: p. 7598.

[226] Bahadar Khan, S., et al., *Photovoltaic Performance of Porphyrin-Based Dye-Sensitized Solar Cells with Binary Ionic Liquid Electrolytes.* Energy Technology, 2020. **8**(6): p. 2000092.

[227] Cao, Y., et al., *11% efficiency solid-state dye-sensitized solar cells with copper(II/I) hole transport materials.* Nature Communications, 2017. **8**(1): p. 15390.

[228] Ravi, S.K., et al., *Optical manipulation of work function contrasts on metal thin films.* Science Advances, 2018. **4**(3): p. eaao6050.

[229] Naghdi, S., G. Sanchez-Arriaga, and K.Y. Rhee, *Tuning the work function of graphene toward application as anode and cathode.* Journal of Alloys and Compounds, 2019. **805**: p. 1117-1134.

[230] Peng, X., et al., *Low Work Function Surface Modifiers for Solution-Processed Electronics: A Review.* Advanced Materials Interfaces, 2018. **5**(10): p. 1701404.

[231] Zhou, Y., et al., *A Universal Method to Produce Low–Work Function Electrodes for Organic Electronics.* Science, 2012. **336**(6079): p. 327.

[232] Ohisa, S., et al., *Doping of Tetraalkylammonium Salts in Polyethylenimine Ethoxylated for Efficient Electron Injection Layers in Solution-Processed Organic Light-Emitting Devices.* ACS Applied Materials & Interfaces, 2019. **11**(28): p. 25351-25357.

[233] Duan, J., et al., *Cationic Polyelectrolytes with Alkylsulfonate Counterions as a Cathode Interface Layer for High-Performance Polymer Solar Cells.* ACS Applied Materials & Interfaces, 2020.

Chapter 4

Ionic Liquids as High-Performance Lubricants and Lubricant Additives

Hong Guo and Patricia Iglesias Victoria

Abstract

Taking into account the environmental awareness and ever-growing restrictive regulations over contamination, the study of new lubricants or lubricant additives with high performance and low toxicity over the traditional lubes to reduce the negative impact on the environment is needed. In this chapter, the current literature on the use of ionic liquids, particularly protic ionic liquids, as high-performance lubricants and lubricant additives to different types of base lubricants are reviewed and described. The relation between ionic liquids structures and their physicochemical properties, such as viscosity, thermal stability, corrosion behavior, biodegradability, and toxicity, is elaborated. Friction reduction and wear protection mechanisms of the ionic liquids are discussed with relation to their molecular structures and physicochemical properties.

Keywords: ionic liquids, friction, wear, tribofilm, additives

1. Introduction

Friction and wear are inescapable problems in mechanical and electromechanical systems, resulting in massive energy losses. Holmberg and Erdemir [1] have estimated that the energy consumption generated by contacting surfaces in mechanical elements is almost 23% of the total energy consumption in the world, where 20% is used to overcome friction and 3% is used to replace worn surfaces. However, energy losses could be reduced by up to 40% through new advances in lubrication, which can save 8.7% of world energy consumption. Especially, the losses by friction could be decreased by using high-performance lubricants, which cannot only result in economic savings but also in important environmental benefits. In addition, the increase in energy prices leads to high demand for improvement of energy efficiency.

Ionic liquids (ILs) are a class of salts composing of bulky organic cations and organic or inorganic anions. Some of the typical IL molecular structures are shown in **Figure 1**. The large molecular size of the ions and their possible delocalized charge contribute to the uncommonly low melting points of ILs, which are below 100 °C. The first IL, ethylammonium nitrate [$(C_2H_5NH_3)NO_3$], reported by Walden in 1914 is found to have a melting point of 12 °C. Since the 1970s, the research of ILs has become increasingly popular and now ILs have been used for

IntechOpen

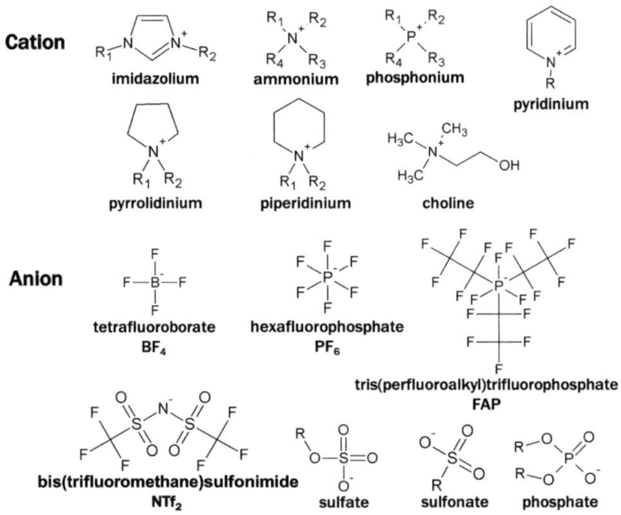

Figure 1.
Typical ionic liquids molecular structures.

various applications such as effective solvent, catalyst, electrolytes in batteries, and carbon and carbon dioxide capturing.

In the tribological field, ILs have shown great potential as advanced lubricants and "tailor-made" lubricating additives since been explored for lubrication in 2001 [2]. ILs have excellent physicochemical properties including low melting point, low flammability, negligible vapor pressure, and high thermal stability that meet the demands of high-performance lubricants. One of the most important characteristics of ILs is that their properties can be tailored by varying the species of the cations and anions, giving rise to numerous families that can be used across different tribological systems. The superiority of ILs in lubrication can be attributed to their inherent polarity, which can make them form stable ordered layers in the liquid state on metal surfaces to prevent them against contact; and that some elements of ILs can react with the substrate materials to generate a tribofilm to protect the substrate from further wear.

ILs can be conventionally categorized into aprotic ionic liquids (AILs) and protic ionic liquids (PILs), based on the nature of the cation present in the combinations [3]. Since most of the studies in lubrication are focused on AILs, many literature reviews have summarized the research efforts of them. Therefore, this chapter covers more about the progress of PILs in lubrication. Firstly, some important physicochemical properties of ILs will be introduced. Secondly, ILs as neat lubricants, specifically as bulk lubricants, thin-film lubricants, and the surface interactions between ILs and contact surfaces will be discussed. The third part will be focused on the ILs as additives in different base lubricants.

2. Physicochemical properties of ionic liquids

ILs are highly tunable by changing the cation structures, anion structures or both to satisfy specific engineering and manufacturing requirements. Therefore, to understand the relationship between the chemical structures of ILs and their physicochemical properties, as well as the tribological properties, becomes crucial to the molecular design of the more effective ILs. Their physicochemical properties can be easily influenced by combining different types of cations and anions or varying their alkyl chain lengths.

2.1 Viscosity and thermal stability

The viscosity behavior will affect the load-carrying capacity of ILs, as well as their formation of boundary lubricating films. In addition, the thermal stability of an IL is also a prerequisite for being used in various tribological systems. Particularly, an outstanding thermal stability would contribute to its application in high-temperature environments. Since thermogravimetric analysis (TGA) has been employed in most of the studies to characterize the thermal stability of ILs, the viscosity and onset thermal decomposition temperature (T_d) of some ILs obtained using this method have been summarized in **Table 1**. In general, the molecular structure modification of the cation or the anion will affect the IL's viscosity and thermal stability. ILs having symmetric cations with long alkyl side chains are found to have high viscosity [16–18], which is attributed to the closer packing and enhanced van der Waals interactions between the long alkyl chains. Particularly, the branched ILs are reported to possess higher viscosity than the linear ones [19]. In addition, ILs having high molar mass and ion-interactions such as hydrogen bonds in their molecular structures will get high viscosity [13, 20]. For instance, an imidazolium-based IL with a hydroxyl group (-OH) grafted into the N-1 position of its cation obtained an increase in viscosity, which is attributed to the increased hydrogen bond interactions and the resulting higher molar mass. In the study of Guo et al. [15], the viscosity of the hydroxylammonium PIL is highly dominated by the hydrogen bond interactions among its molecules instead of its molar mass.

The thermal stability of an IL is also closely related to its cation and anion, as can be seen in **Table 1**. Generally, when pairing with the same anion, the imidazolium-based ILs have a higher thermal stability than the tetraalkylphosphonium-based and the tetraalkylammonium-based ILs [4, 5, 9]. And the imidazolium-based ILs are reported to have a higher thermal stability when their cations have a smaller alkyl chain [21]. In contrast to cations, the anions have more impacts on the thermal stability of ILs. For example, an alkylammonium PIL derived from a stronger acid tends to have a higher thermal stability [22]. Recently, Fadeeva et al. [23] reported that the thermal stability of the alkylimidazolium-based PILs is mainly determined by the anions nature instead of the cation structure. However, with the same triflate anion, the PILs having the cation with larger size and branched chain structure would get a higher thermal stability [24, 25]. In several studies [13–15], the hydroxylammonium PILs with carboxylate anions were found to have low thermal stability. The reason underlaying this phenomenon is related to the reversal proton transfer, leading to the presence of free acid and ethanolamine [15]. Considering the long-term practical applications of ILs in lubrication, the characterization of their long-term thermal stability, through isothermal TGA, should also be concerned [21].

2.2 Corrosion

Corrosion for most lubricants, such as water-based lubricants, is a complex problem that needed to be solved. The corrosivity of neat ILs or ILs additives are usually evaluated by means of immersion corrosion test, in which the testing specimen, such as copper [26–29], steel [28, 30], or cast iron [29] will be immersed into (or cover the metal surface with) neat ILs or IL containing lubricants. In addition, electrochemical corrosion tests will also be conducted to examine the anticorrosion properties of ILs as well as study the corrosion mechanisms [29, 31]. Through the

Cation	Anion	Viscosity (25 °C)	Viscosity (40 °C)	Viscosity (100 °C)	Viscosity Index	Thermal Stability T_d (°C)	Ref
[C_4C_1im]	NTf_2	60.7	50.1	46	—	413.85	[4]
[C_4im]		129.1	52.1	46.2	—	382.85	
N_{4441}	NTf_2	~480	~190	~20	—	360	[5]
N_{8886}	BScB	$\sim 190^{\Psi}$	~110	~20	—	—	[6]
N_{8888}		$\sim 690^{\Psi}$	~370	~50	—	—	
N_{88810}		$\sim 200^{\Psi}$	~120	~20	—	—	
N_{88812}		$\sim 110^{\Psi}$	~80	~20	—	—	
N_{8881}	$C_{6:0}$	2752.9^{Ψ}	1284.9	56.5	92		[7] [8]
	$C_{8:0}$	2410.3^{Ψ}	1121.2	48.6	85	175.2	
	$C_{12:0}$	1475.8^{Ψ}	715.7	36.9	85	175.1	
	$C_{16:0}$	1188.3^{Ψ}	596.3	37.4	99	183.3	
	$C_{18:0}$	1033.2^{Ψ}	524.2	35.1	102		
	$C_{18:1}$	1234.7^{Ψ}	627.2	39.3	101		
P_{66614}	$C_{10:0}$	141800*	16278*	16.91*	—	268.65	[9]
P_{66614}	DEHP	1031^	418.4	49.2	—	~300	[10]
P_{66614}	$(iC8)_2PO_2$	1204.1	528.93	55.06	169	300.73	[11]
P_{66614}	BEHP	1156	528.05	59	181	293.48	
P_{66614}	NTf_2	277.26	123.49	16.15	140	417.3	
P_{44414}	DBS	4423.1	1355.2	62.34	98	333.01	
P_{4442}	DEP	451.39	171.53	14.81	83	304.1	
a	Citrate	321691*	65097*	941.21*	—	191.2	[12]
a	Succinate	32798*	8123.5*	167.1*	—	178.6	[13]

DOI: http://dx.doi.org/10.5772/intechopen.96428

Cation	Anion	Viscosity (25 °C)	Viscosity (40 °C)	Viscosity (100 °C)	Viscosity Index	Thermal Stability T_d (°C)	Ref
b	Formiate	16.13	—	—	—	150	[14]
	Pentanoate	1333.2	—	—	—	122	
b	Hexanoate#	8943.8*	2405.56*	59.4*	—	177.48	[15]
c	Hexanoate#	926.86*	301.58*	14.72*	—	174.82	
d	Hexanoate#	61.74*	27.74*	3.71*	—	141.44	

Note: a- $NH_2((CH_2)_2OH)_2$; b- $NH_3((CH_2)_2OH)$; c- $NH_2CH_3(CH_2)_2OH)$; d- $NH(CH_3)_2(CH_2)_2OH)$.
#2-ethylhexanoate.
**Dynamic viscosity (cP).*
^kinetic viscosity at 23 °C.
Ψkinetic viscosity at 30 °C; and the numbers in imidazolium $C_xC_x$im, ammonium $N_{x,x,x,x}$, and phosphonium $P_{x,x,x,x}$ represent the alkyl chain length.

Table 1.
Kinetic viscosity and thermal stability of some ILs.

two methods, the anticorrosion performance of four hydroxylammonium phosphate PILs additives was investigated in [28]. Compared to the reference sample immersed in water, the use of neat PILs significantly improved the corrosion resistance of the copper and iron sheets. What is more, the results from the electrochemical test further demonstrated the excellent corrosion inhibition of PILs additives and their attributes of anodic corrosion inhibitors. A protective film generated by the adsorption of PIL molecules, particularly the hydrophilic functional group on steel surface is attributed to anticorrosion performance. Among the four PILs, 2-hydroxypropylammonium di-(2-ethylhexyl) phosphate ($TEOAP_6$) was found to have the best corrosion resistance, and PIL's corrosion inhibition efficiency was related to its functional groups.

In the study of Ma et al. [26], the immersion corrosion test was employed to evaluate the anticorrosion property of PAO when ILs additives were added in different concentrations. Along with ILs, an additive containing Sulfur element was also added to PAO in 1%. It is noted that the addition of only 0.1% ILs can greatly reduce the corrosion tendency of a lubricant, and IL concentration (0.25–0.1%) just slightly affected the corrosion-inhibiting performance. Recently, a PIL 2-hydroxyethylammonium oleate [32], was proved to be an effective corrosion inhibitor for aluminum 1100 in neutral sodium chloride solution. The PIL adsorption layer on the aluminum substrate surface was pointed out to inhibit the diffusion of chloride anions during the electrochemical measurement, where it can also provide a corrosion protection at high chloride concentration for 72 hours.

2.3 Biodegradability and toxicity

In terms of the prospective large-scale industrial applications of ILs, it is crucial to examine their biodegradability and toxicity to control the discharge of IL-involved solvents or lubricants, minimizing the environmental damage. To understand the correlation between the molecular structure and biodegradability and toxicity of ILs is essential to design green IL lubricants. Nowadays, many experimental studies [7, 33] have been conducted to evaluate the environmental impact of ILs. In addition, computational approaches [34–36] are also employed to assess their toxicity. In general, cations and anions do have an influence on the ILs toxicity particularly, cations have a greater impact than anions. ILs having longer alkyl chain length and more branched-chain groups on their cations tend to be more toxic [37]. However, some anions containing fluorine in their structure will cause an increase in toxicity of their corresponding ILs. For example, although the hydroxylammonium and imidazolium cations were evaluated to be less toxic, the toxicity of their ILs increased drastically once NTf_2 was incorporated as the anion [38]. Regarding the biodegradability, the IL components 1–Butyl–3–methylimidazolium (Bmim) and bis(trifluoromethanesulfonyl) imide (NTf_2) were reported to be non-biodegradable even at low concentration (10 mg/L), while the N,N,N–trimethylethanolammonium (Choline) and acetate (Ac) could be completely degraded with a concentration up to 50 mg/L [38].

In the study of Tzani et al. [39], the biodegradability of a series of carboxylate-PILs were examined and proved to be relevant to the alkyl chain length of the anions. PILs having anions with long alkyl chain length were found to get a decreased biodegradability, except for the one that had an alicyclic ring in its anion, showing an enhanced biodegradability. Lately, Viesca et al. [33] characterized the biodegradability and bacteria toxicity of six PILs derived from alkylhydroxylamine. Owing to the presence of benzenesulfonate aromatic group in anions, the sulfonate-PILs were reported to be less biodegradable compared to the hexanoate-PILs. Regarding the bacterial toxicity behavior, even the hexanoate-PILs exhibited a better environmental impact, all of them were mild toxic to Vibrio fischeri. Nevertheless, all these PILs were found to outperform the traditional lubricant additive ZDDP concerning the biodegradability and toxicity performance.

3. Ionic liquids as lubricants

The tribological behavior of ILs as lubricants have been typically evaluated through laboratory bench tests using various macroscopic tribometers, such as the Optimol SRV series tribometers, mini-traction machines, Microtest pin-on-disk tribometer, Plint TE77 high-frequency reciprocating rigs, etc. In addition, the atomic force microscope (AFM) and surface force apparatus (SFA) are usually applied to investigate the nanotribological performance of lubricants. The two main factors, coefficient of friction (COF), and wear volume (or wear rate) of the rubbing materials are normally used to evaluate and compare the lubricating ability and anti-wear performance of IL lubricants.

3.1 Ionic liquids as neat lubricants

Since Ye et al. [2] initiated the study of ILs in lubrication in 2001, the studies about ILs as neat lubricants for various contact systems such as steel-steel contact [15], steel-ceramic contact [40], and steel-aluminum contact [14] have received considerable attention. **Table 2** summarizes some recent studies of ILs as neat lubricants. Compared to AILs, the use of PILs as neat lubricants has gained more attention than before, owing to their low cost and facile synthesis process.

In Khan et al.'s research, two phosphonium-based PILs with different alkyl chain length in the anions were tested as neat lubricants under steel-steel contact, and a synthetic oil PEG 200 was used as a reference [41]. Since the fatty acid anions of PILs have a better affinity to steel surfaces, the use of PILs showed a significant friction reduction with respect to PEG 200. The tribological performance of the two PILs were found to be determined by the alkyl chain length of their anions and their viscosity, where a PIL with a shorter anion chain length and lower viscosity led to a lower friction coefficient but more material loss. While the results may be inverse once the experiment conditions are changed or other PILs are used. In the study of Vega et al. [14], the effect of anion chain length on the friction and wear behavior of ammonium-based PILs was investigated under steel-aluminum contact. The results revealed that increasing the anion chain length will improve the lubricating ability of PIL with a low friction coefficient. From another study of Vega et al. [42], three oleic-acid derived ammonium-based PILs were evaluated as lubricants in alumina-aluminum contact. In addition to the low friction coefficient, the use of PILs yielded an important wear reduction (98%) compared to the dry condition. Lately, the hexanoate-based PILs were also found to greatly reduce the wear of steel with respect to mineral oil as well as a commercial oil [15]. Tribofilms were detected on the worn steel disks when PILs were used to protect the steel against severe wear.

In addition to the above-mentioned bulk lubricants, ILs can also be employed in the form of thin layers for lubricating micro/nano electromechanical systems (MEMS/NEMS). For example, in Bermúdez's group [40], a PIL - di[bis(2-hydroxyethyl)ammonium] succinate thin layer was created on a steel substrate surface by evaporating water from the PIL + Water mixture, where the PIL thin layer extremely reduced the wear rate of steel compared to the bulk neat PIL.

Cation	Anion	Tribo-pair	Contact mode	Load (N)	COF	Ref
b	Hexanoate#	Steel/steel	Ball-on-flat	3	0.038	[15]
c					0.032	
d					0.58	
b	Formiate	Steel/Al	Ball-on-plate	0.5	0.35 ± 0.12	[14]
	Pentanoate				0.14 ± 0.026	
P_{888H}	Caprylate	Steel/steel	Four-ball	392	0.038	[41]
	Oleate				0.044	
b	Oleate	Alumina/Al	Ball-on-plate	0.5	0.11–0.15	[42]
c					0.12–0.15	
a					0.12–0.076	
$[C_4C_1im]$	BF_4	Steel/steel	Ball-on-disk	20	~0.066	[43]
	PF_6				~0.0.8	
	BF_4			40	0.06	
	PF_6				0.07	
	BF_4			60	0.06	
	PF_6				~0.075	
a	Succinate	Sapphire/steel	Pin-on-disk	0.98	0.119	[40]
N_{HHH10}	Oleate	Steel/steel	Ball-on-plate	4	$0.048^{\delta}/0.054^{\lambda}$	[27]
IL-TO					$0.069^{\delta}/0.056^{\lambda}$	
a					$0.063^{\delta}/0.066^{\lambda}$	
P_{66614}	$(iC_8)_2PO_2$				$0.069^{\delta}/0.093^{\lambda}$	

Note: a- $NH_2((CH_2)_2OH)_2$; b- $NH_3((CH_2)_2OH)$; c- $NH_2CH_3((CH_2)_2OH)$; d- $NH(CH_3)_2(CH_2)_2OH)$.
#2-ethylhexanoate.
$^{\delta}$coefficient of friction at 30 °C.
$^{\lambda}$coefficient of friction at 80 °C; and the numbers in imidazolium C_xC_xim, ammonium N_{HHHx}, and phosphonium $P_{x,x,x,x}$ represent the alkyl chain length.

Table 2.
Tribological results of some ILs as neat lubricants (2017–2020).

3.2 Surface interactions

As shown in **Figure 2**, it has been widely accepted that when neat ILs or IL additives are introduced between the contacting work pairs, the IL molecules tend to adsorb onto the workpiece surfaces physically or/and chemically and form an ordered boundary lubricating film to protect the moving components from direct contact, leading to low friction. During the sliding frictional process, a protective tribofilm will be subsequently generated on top of the substrate by means of the tribochemical reactions between ILs or their decomposition products and the contacting metal surfaces to reduce mechanical wear.

Although the process of forming the adsorbed boundary lubricating film is still not clear, the IL-adsorption film has been verified through electrical contact resistance (ECR) measurement by Viesca et al. [44]. The results showed that the IL-additive ($[C_6C_1im][BF_4]$) outpaced the base oil to form a boundary film on the metal surface. The generation of the IL-tribofilm has been demonstrated on various material surfaces [45–47]. But the results from most of the work are relied on the post-analysis of the worn surfaces by employing Scanning Electron Microscopy

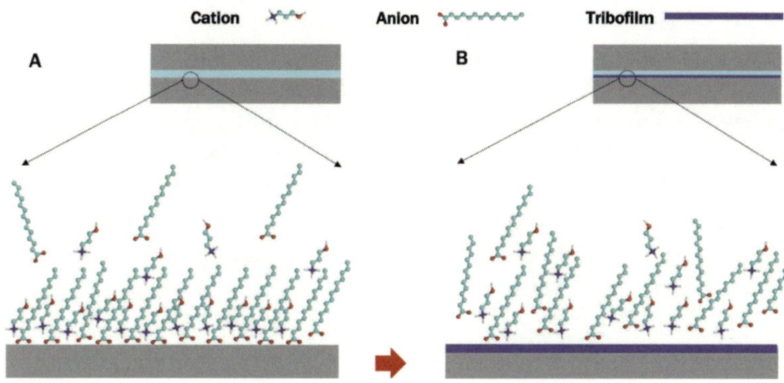

Figure 2.
Schematic diagram of (A) ILs boundary lubricating film, and (B) IL-induced tribofilm on the metal surface.

(SEM), Transmission Electron Microscopy (TEM), Energy-dispersive X-ray Spectroscopy (EDS), Auger Electron Spectroscopy (AES), X-ray Photoelectron Spectroscopy (XPS), Raman Spectroscopy, etc. From the previous research [48], IL decomposition has been demonstrated during the sliding process, but only the anion was found to react with or adsorbed on the steel surface. Particularly, the IL undergoing facile decomposition would interact rapidly with the sliding surface, leading to a low friction coefficient. So the thermal stability of IL can be considered as an index for evaluating the tribo-decomposition behavior on nascent substrate surfaces [49].

Up to now, the characterization of the IL-induced tribofilm thickness, composition, and structure have been intensively investigated. For instance, when phosphonium-phosphate ILs were introduced to the base oils with a small amount (1.04 wt.%), an amorphous-nanocrystalline tribofilm with a 10–200 nm-thick was probed on the worn cast iron surface by TEM, EDS, and electron diffraction [17]. Furthermore, the participation of wear debris in the IL-tribofilm growth was proposed and demonstrated recently by Qu et al. [45, 46] through Atom Probe Tomography (APT) and Scanning Transmission Electron Microscopy (STEM) characterization.

In addition, tribofilm mechanical properties, such as hardness and resistance-to-plastic-deformation (P/S^2), have also been investigated through nanoindentation measurements [50, 51]. The results revealed that only P/S^2 had a correlation with the friction and wear performance, in which a small P/S^2 value corresponded to a low friction and wear.

Regarding the growth mechanism of IL-induced tribofilm, a more precise *in situ* characterization is highly desirable in spite of many characterization approaches and spectroscopy techniques have been employed so far. At the same time, the application of the computational methods, such as molecular dynamic simulation, would help to elucidate the generation process of the boundary lubricating film.

4. Ionic liquids as lubricant additives

Limited to the high cost of being used as neat lubricants (particularly when AILs are used), ILs as additives have gained more and more research attention in recent years. Their highly tunable molecular structures and physicochemical properties make ILs suitable to be added to base lubricants with different nature (polar or nonpolar), such as ester, polyethylene glycol (PEG), PAO, mineral oils (MO), grease, and water-based lubricants.

Until now, ILs have been tested as friction-reducing additives, anti-wear additives, or extreme-pressure additives in many research articles. The tribological performance of IL as additives to non-polar, polar, and water-based lubricants have been summarized in **Tables 3–5**, respectively. Different from the traditional friction modifiers, ILs can be strongly adsorbed to the sliding surfaces and generate a resilient boundary lubricating film, leading to important reduction of friction and wear. Some active-elements containing ILs are easy to chemically react with the rubbing surfaces and create an effective tribofilm on top of the workpieces to prevent it against wear or extreme pressure.

4.1 Ionic liquids as oil additives

Due to the inherent polarity, the solubility of ILs in oils is a complicated issue. Most imidazolium-ILs are insoluble in the non-polar synthetic oils and mineral oils. So they are always used as lubricant additives in in very low concentrations or in oil-IL emulsions. In 2012, the fully oil-soluble phosphonium-based ILs [$P_{6,6,6,14}$][DEHP] and [$P_{6,6,6,14}$][BTMPP] were explored [74, 75]. These three-dimensional ILs have quaternary structures for both the cations and anions with long alkyl chains, giving rise to a high steric hindrance to screen the ions charge. Inspired by this, ILs having quaternary ammonium and phosphonium cations and halogen-free anions, such as phosphate, sulfonate, orthoborate, and carboxylate have been synthesized and tested as additives to the base oils [76]. Generally, larger cation sizes lead to higher solubilities of IL in nonpolar oils. In addition, ILs having symmetric cations would outperform the ones with asymmetric cations in wear reduction, and the symmetric-cation ILs are hypothesized to have a better mobility in the base oil to interact with metal surfaces and form protective boundary lubricating film [17]. In addition, some phosphonium-based ILs have been examined to show synergistic interactions with traditional additives ZDDP in hydrocarbon oils [77], or GTL base oil [78] to improve the wear resistance of oils.

In contrast to nonpolar oils, ILs have much better solubility in some polar oils, such as PEG200, in which [$C_6C_1C_1$im][NTf2] can be dissolved up to 40 wt.%. Taher et al. [79] studied the lubricating properties of halogen-free ILs pyrrolidinium bis (mandelato)borate (hf-BILs) as additives to PEG200 in steel-steel contact. The addition of 3 wt.% of hf-BILs in the base oil reduced friction and wear significantly compared to PEG200 and 5 W40 engine oil. It is noted that shorten the length of the longest alkyl chain in this IL cation will improve the friction reduction and wear resistance of the IL-blends under same working conditions.

Recently, Guo et al. [57, 80] examined the tribological properties of three hydroxylammonium hexanoate PIL additives to a nonpolar mineral oil and a polar biodegradable oil. The impact of PILs ionicity and hydrogen bonding on the friction and wear performance was discussed. The results revealed that all PILs improved the lubricity and wear resistance of the biodegradable oil under steel-steel, particularly, the one with the lowest ionicity obtained the least material loss. While, when used as additives to the mineral oil, the three PILs behaved slightly different between steel-steel and steel-aluminum contact. The use of any PIL improved the mineral oil lubricity and wear resistance under both contacts, but PILs had quite different friction behaviors in steel-Al that the one with the highest ionicity presented the best friction.

4.2 Ionic liquids in water-based lubricant

Water or water-based lubricants can effectively reduce the temperature and clean the contaminants from surface contacts, which leads to a better working

Cation	Anion	Base Oil	Content (wt.%)	Tribo-pair	Contact mode	Load (N)	COF	Ref
$N_{121212P}$	DOSS	PAO10	1–3	Steel/steel	Ball-on-disk	50	0.1–0.12	[52]
	Laurate							
	DOSS					200	0.09–0.11	
	Laurate							
N_{888H}	DEHP	PAO4	0.87	Steel/ bronze	Ball-on-flat	20	0.09	[53]
a	Succinate	PAO40	1	Steel/steel	Ball-on-flat	3	0.085	[13]
	Citrate						0.065	
	Succinate					4	0.075	
	Citrate						0.059	
a	Citrate	MO	1	Steel/steel	Ball-on-flat	2	0.1	[12]
b	Hexanoate[#]	MO	1	Steel/steel	Ball-on-flat	3	0.084	[80]
c							0.085	
d							0.086	
a				Steel/Al			0.101	
b							0.061	
c							0.046	
P_{8888}	DEHP	GTL4	1.04	Steel/iron	Ball-on-flat	100	0.115	[46]
P_{66614}	$(iC8)_2PO_2$	MO	1	Steel/steel	Ball-on-flat	2	0.09	[54]
	NTf_2						0.127	
P_{66614}	DEHP	PAO4	1	Steel/ OD-Ti64	Ball-on-flat	100	0.05	[55]
	Stearate		0.99			100	~0.08	
	BTMPP		0.99			100	~0.07	
N_{888H}	DEHP		0.87			100	~0.06	

Cation	Anion	Base Oil	Content (wt.%)	Tribo-pair	Contact mode	Load (N)	COF	Ref
P_{88816}	DOSS	500SN	1–4	Steel/Al	Ball-on-disk	100	0.11–0.125	[47]
	DOSS			Steel/Mg		100	0.1–0.115	
P_{888P}	DOSS			Steel/Al		100	0.11–0.125	
	DOSS			Steel/Mg		100	0.1–0.115	
P_{66614}	$(iC8)_2PO_2$	PAO4	0.5	Steel/steel	Ball-on-plate	60	0.015–0.02	[56]
	BEHP						~0.017	
	$(iC8)_2PO_2$		1				~0.015	
	BEHP						~0.014	
TTAOA	DEHP	PAO10	0.25–1	Steel/steel	Ball-on-disk	200	0.107–0.11	[26]
TTADO								

Note: a- $NH_2((CH_2)_2OH)_2$; b- $NH_3((CH_2)_2OH)$; c- $NH_2CH_3(CH_2)_2OH)$; d- $NH(CH_3)_2(CH_2)_2OH)$.
#2-ethylhexanoate; and the numbers in ammonium $N_{X,X,X,X}$, and phosphonium $P_{X,X,X,X}$ represent the alkyl chain length.

Table 3.
Tribological results of ILs as additives for non-polar oils (2017–2020).

Cation	Anion	Base Oil	Content (wt.%)	Tribo-pair	Contact mode	Load (N)	COF	Ref
b	Hexanoate[#]	BO	1/2	Steel/steel	Ball-on-flat	3	0.07/0.05	[57]
c							0.052/0.035	
d							0.035/0.059	
N_{1888}	NTf_2	Diester	1.25–5	Steel/steel	Ball-on-disk	40	0.069–0.071	[58]
						80	0.068–0.07	
						120	0.068–0.07	
a	Citrate	BO	1	Titanium/ ceramic	Ball-on-flat	2	~0.12	[59]
N_{6666}	Octanoate	POE	0.5/2	Steel/steel	Ball-on-plate	30	0.062/0.059	[60]
						50	0.065/0.061	
	Palmitate					30	0.066/0.062	
						50	0.066/0.063	
P_{66614}	Stearate	PETO	4	Steel/steel	Ball-on-disk	200	0.105–0.12	[61]
		TMPTO					0.12–0.16	
	Oleate	PETO					0.105–0.15	
		TMPTO					0.12–0.16	
	$(C_1)_2S_2PO_2$	PETO					0.075	
		TMPTO					0.09	
P_{66614}	$(iC8)_2PO_2$	MJO	1–10	Steel/steel	Four-ball	392	0.05–0.075	[62]
N_{1888}	NTf_2							
P_{66614}	SiSO	SQL	1	Steel/steel	Ball-on-disk	600	0.04–0.07	[63]

Cation	Anion	Base Oil	Content (wt.%)	Tribo-pair	Contact mode	Load (N)	COF	Ref
P_{66614}	$(iC8)_2PO_2$	BO	1	Steel/steel	Pin-on-disk	4.9	0.125–0.22	[64]
	NTf_2						0.12–0.28	
N_{4444}	Sulphate	PEG 200	1.5	Steel/steel	Four-ball	392	0.12	[65]
N_{8888}							0.1	
P_{4444}							0.13	
P_{8888}							0.13	

Note: a- $NH_2((CH_2)_2OH)_2$; b- $NH_3((CH_2)_2OH)$; c- $NH_2CH_3(CH_2)_2OH)$; d- $NH(CH_3)_2(CH_2)_2OH)$.
#2-ethylhexanoate; and the numbers in ammonium $N_{X,X,X,X}$, and phosphonium $P_{X,X,X,X}$ represent the alkyl chain length.

Table 4.
Tribological results of ILs as additives for polar oils (2017–2020).

conditions and increase the machine lifetime. Since the high volatile characteristic and high freezing point of water-based lubricants, they are preferable in some specific industrial applications such as cutting and machining. Recent studies about IL additives in water are summarized in **Table 5.**

In the study of Wang et al. [81], N-(3-(diethoxyphosphoryl)propyl)-N,N-dimethyloctadecan-1-ammonium bromide (NP) was investigated as water additive in a steel-steel contact. A lower friction and wear rate, and excellent extreme-pressure and abrasion resistance were obtained compared to an oil-based lubricant. The superior tribological property was attributed to the physical adsorption of ILs on the steel surfaces and the formation of a protective film due to the tribo-chemical reactions between NP and sliding surfaces.

Bermudez's team [82] reported that water containing 1 wt.% PIL (2-hydroxyethylammonium) succinate (MSu) could reduce the running-in period when lubricating the sapphire-stainless steel contact. It is also noted that a thin PIL boundary film was found on the steel surface once the base water evaporated, leading to an extremely low minimum friction coefficient of 0.0001. In addition, another PIL additive, di[bis(2-hydroxyethyl)ammonium] succinate (DSu) was also investigated under sapphire-stainless steel [40]. The results showed that although the use of 1 wt.% DSu + Water caused a higher running-in friction coefficient compared to that of neat DSu, PIL-mixture received a comparable anti-wear behavior with regards to the neat Dsu, and even got a slightly smaller wear rate of 1.83×10^{-5} mm^3/m.

4.3 Ionic liquids and nanoscale additives

Nanomaterials, such as nanoparticles (NPs), graphene, and carbon nanotubes (CNTs), have been regarded as attractive solid lubricants which can be applied as lubricant additives and components for coatings to achieve good lubricity or super-lubricity. In [83], the magnesium silicate hydroxide–based nanoparticles have been studied and proved to be effective anti-wear additives, where the excellent tribological properties can be generally ascribed to the grinding, rolling, filling effects and the tribofilm formation.

However, the poor dispersion and low solubility of nanomaterials in the base lubricants limit their long-term practical applications. Therefore, the nanomaterial surface functionalization becomes necessary to their lubrication performance. The use of an oil-soluble PIL with long-alkyl-chain to incorporate the copper oxide nanoparticles as additives to a base oil PAO was firstly reported in [84]. In this study, the PIL was employed to improve the dispersion of the copper oxide NPs, where the hybrid PIL-NPs additives exhibited an enhanced oil-load capacity and a better anti-wear performance compared to that just using copper nanoparticles as additives. Recently, the friction behavior and wear performance of diamond and ZnO NPs stabilized by trihexyltetradecylphosphonium bis (2, 4, 4-trimethylpentyl) phosphinate were investigated in a steel-ceramic contact [85]. It was found that nanoparticles mixed with IL caused a higher friction coefficient with respect to only IL was used as additive to the gear base oil, where the nanoparticles were regarded as to wear out the film formed by the IL. While the use of diamond/ZnO nanoparticles with IL obtained a smaller wear volume of the ceramic ball compared to that of IL. Particularly, both the hybrid IL-nanoadditives showed effective anti-scuffing properties which revealed their potential to be extreme pressure additives.

Cation	Anion	Content (wt.%)	Tribo-pair	Contact mode	Load (N)	COF	Ref
C_6C_1im	Ibu	2	Steel/steel	Ball-on-disk	100	0.135	[29]
C_8C_1im	Ibu	2	Steel/steel	Ball-on-disk	100	0.127	
N_{16111}^{β}	P1	0.4	Steel/steel	Four-ball	392	0.075	[66]
N_{16111}^{β}	P2	0.4	Steel/steel	Four-ball	392	0.085	
N_{4444}	BTA	0.03^{ε}	Steel/steel	Ball-on-plate	100	0.125	[67]
P_{4444}	BTA	0.03^{ε}	Steel/steel	Ball-on-plate	100	0.125	
NP_{1611}^{χ}	GAS	0.5	Steel/steel	Ball-on-disk	100	0.110	[68]
NP_{1611}^{χ}	AK	0.5	Steel/steel	Ball-on-disk	100	0.095	
$NP_{1611}^{\phi 1}$	Br	0.5	Steel/steel	Ball-on-disk	100	0.075	[69]
$NP_{1611}^{\phi 2}$	Br	0.5	Steel/steel	Ball-on-disk	100	0.100	
b	Stearate	1	Steel/sapphire	Pin-on-disk	1	0.129	[70]
a						0.117	
a	Palmitate					0.107	
b^{φ}	Ricinoleate	1	Steel/steel	Ball-on-disk	125	0.14–0.15	[71]
	MBT					0.18–0.2	
a	Citrate	1	Al/tungsten carbide	Pin-on-disk	2.94	~0.45	[9]
P_{66614}	NTf_2					~0.57	
P_{66614}	Decanoate					~0.18	

Cation	Anion	Content (wt.%)	Tribo-pair	Contact mode	Load (N)	COF	Ref
a	Oleate	1	Steel/steel	Ball-on-plate	2	0.078	[72]
d						0.066	
a			Alumina/steel		2	0.079	
d						0.09	
a			Steel/steel		4	0.091	
d						0.076	
a			Alumina/steel		4	0.084	
d						0.094	
c^{φ}	Ricinoleate	1	Steel-steel	Ball-on-disk	125	~0.12	[73]
	Phosphate					~0.2	

Note: The base lubricant of β is water-glycol.
a- $NH_2((CH_2)_2OH)_2$; b- $NH_3((CH_2)_2OH)$; c- $NH_2CH_3(CH_2)_2OH)$; d- $NH_3(CH_3)_2CCH_2C(CH_3)_3$; and the numbers in imidazolium C_xC_xim; ammonium $N_{X,X,X,X}$, and phosphonium $P_{X,X,X,X}$ represent the alkyl chain length.
$^{\chi}$water-diethylene glycol.
$^{\phi1}$water-sodium D-gluconate.
$^{\phi2}$water-triethanolamine.
$^{\varphi}$water-glycerol.
$^{\varepsilon}$mol/L.

Table 5.
Tribological results of ILs as additives for water-based lubricants (2017–2020).

5. Conclusions

As the aforementioned excellent physicochemical properties and friction and wear performance, ILs not only can be used as neat lubricants, friction-reducing additives, anti-wear additives, extreme pressure additives, but can also be used as corrosion inhibitors. Although IL corrosion inhibitors have been evaluated on many ferrous metals and alloys, their study on non-ferrous metals, such as aluminum is extremely limited, which is worthwhile to discuss. Meanwhile, the relationship between the outstanding corrosion inhibition and high performance of lubrication should be explored, when ILs are used as lubricants and lubricant additives.

Additionally, enormous literature has revealed that the adsorption of the ILs on the metallic surfaces and the tribo-chemical reactions between the active elements of ILs and the surfaces effectively improved the tribological performances of different contacts. However, the adsorption mechanism and tribofilm growth mechanism of ILs are still not clear, and the application of ILs in the tribology field, especially for PILs, should be further explored owing to its efficiency and green nature.

Acknowledgements

Hong Guo wants to express her gratitude to the Gleason Doctoral Fellowship from the Gleason Corporation.

Conflict of interest

The authors declare no conflict of interest.

Acronyms and abbreviations

$[NTf_2]$	bis(trifluoromethylsulfonyl)amide
BScB	bis(salicylato)borates
$C_{6:0}$	hexanoate
$C_{8:0}$	octanoate
$C_{10:0}$	decanoate
$C_{12:0}$	laurate
$C_{16:0}$	palmitate
$C_{18:0}$	stearate
$C_{18:1}$	oleate
$(iC8)_2PO_2$/ BTMPP	bis(2,4,4-trimethylpentyl)phosphinate
BEHP/DEHP	bis(2-ethylhexyl)phosphate
DBS	dodecylbenzenesulfonate
DEP	diethylphosphate
BF_4	tetrafluoroborate
PF_6	hexafluorophosphate
IL-TO	t-octhylammonium
DOSS	dioctyl sulfosuccinate
TTAOA	4(or 5)-methyl-benzotriazole-1-ylmethyl)-octadec-9-enyl-ammonium
TTADO	4(or 5)-methyl-benzotriazole-1-ylmethyl)-dioctyl-ammonium

$(C_1)_2S_2PO_2$ - O,O′	diethyldithiophosphate
SiSO – 3	(trimethylsily)propane-1-sulfonate
Ibu	ibuprofen
P1	phosphate
P2	phosphite
BTA	benzotriazole
$[NP_{1611}][GAS]$	*N*-(3-(Diethoxyphosphoryl)propyl)-*N,N*-dimethyloctadecan-1-aminium-2,3,4,5,6-pentahydroxy-hexanoate
$[NP_{1611}][AK]$	*N*-(3-(Diethoxyphosphoryl)propyl)-*N,N*-dimethyloctadecan-1-aminium-6-methyl-4-oxo-4*H*-1,2,3-oxathiazin-3-ide-2,2-dioxide
Br	bromide
MBT	2-mercaptoben- zothiazole
POE	polyol ester
GTL4	gas-to-liquid 4 cSt
PETO	pentaerythritol oleate
TMPTO	trimethylolpropyl trioleate
MJO	modified Jatropha oil
SQL	squalane
PEG200	polyethylene glycol

Author details

Hong Guo* and Patricia Iglesias Victoria
Mechanical Engineering Department, Kate Gleason College of Engineering, Rochester Institute of Technology, Rochester, USA

*Address all correspondence to: hxg6557@rit.edu

References

[1] Holmberg, K., Erdemir, A.: Influence of tribology on global energy consumption, costs and emissions. Friction. 5, 263–284 (2017). doi:10.1007/s40544-017-0183-5

[2] Ye, C., Liu, W., Chen, Y., Yu, L.: Room-temperature ionic liquids: a novel versatile lubricant. Chem. Commun. (Camb). 2244–2245 (2001). doi: 10.1039/B106935G

[3] Sharma, G., Singh, D., Gardas, R.L.: Effect of Fluorinated Anion on the Physicochemical, Rheological and Solvatochromic Properties of Protic and Aprotic Ionic Liquids: Experimental and Computational Study. ChemistrySelect. 2, 11653–11658 (2017). doi:10.1002/slct.201701985

[4] Shang, D., Zhang, X., Zeng, S., Jiang, K., Gao, H., Dong, H., Yang, Q., Zhang, S.: Protic ionic liquid [Bim][NTf2] with strong hydrogen bond donating ability for highly efficient ammonia absorption. Green Chem. 19, 937–945 (2017). doi:10.1039/c6gc03026b

[5] González, R., Ramos, D., Blanco, D., Fernández-González, A., Viesca, J.L., Hadfield, M., Hernández Battez, A.: Tribological performance of tributylmethylammonium bis (trifluoromethylsulfonyl)amide as neat lubricant and as an additive in a polar oil. Friction. 7, 282–288 (2019). doi: 10.1007/s40544-018-0231-9

[6] Gusain, R., Bakshi, P.S., Panda, S., Sharma, O.P., Gardas, R., Khatri, O.P.: Physicochemical and tribophysical properties of trioctylalkylammonium bis (salicylato)borate (N888n-BScB) ionic liquids: Effect of alkyl chain length. Phys. Chem. Chem. Phys. 19, 6433–6442 (2017). doi:10.1039/c6cp05990b

[7] Oulego, P., Faes, J., González, R., Viesca, J.L., Blanco, D., Battez, A.H.: Relationships between the physical properties and biodegradability and bacteria toxicity of fatty acid-based ionic liquids. J. Mol. Liq. 292, (2019). doi:10.1016/j.molliq.2019.111451

[8] Battez, A.H., Rivera, N., Blanco, D., Oulego, P., Viesca, J.L., González, R.: Physicochemical, traction and tribofilm formation properties of three octanoate-, laurate- and palmitate-anion based ionic liquids. J. Mol. Liq. 284, 639–646 (2019). doi:10.1016/j.molliq.2019.04.050

[9] Del Sol, I., Gámez, A.J., Rivero, A., Iglesias, P.: Tribological performance of ionic liquids as additives of water-based cutting fluids. Wear. 426–427, 845–852 (2019). doi:10.1016/j.wear.2019.01.109

[10] Qu, J., Luo, H., Chi, M., Ma, C., Blau, P.J., Dai, S., Viola, M.B.: Comparison of an oil-miscible ionic liquid and ZDDP as a lubricant anti-wear additive. Tribol. Int. 71, 88–97 (2014). doi:10.1016/j.triboint.2013.11.010

[11] Hernández Battez, A., Bartolomé, M., Blanco, D., Viesca, J.L., Fernández-González, A., González, R.: Phosphonium cation-based ionic liquids as neat lubricants: Physicochemical and tribological performance. Tribol. Int. 95, 118–131 (2016). doi:10.1016/j.triboint.2015.11.015

[12] Guo, H., Iglesias, P.: Tribological behavior of ammonium-based protic ionic liquid as lubricant additive. Friction. 9, 169–178 (2021). doi:10.1007/s40544-020-0378-z

[13] Guo, H., Iglesias, P.: Tribological Properties of Ammonium Protic Ionic Liquids As Additives in Polyalphaolefin for Steel-Steel Contact. In: Volume 12: Advanced Materials: Design, Processing, Characterization, and Applications. American Society of Mechanical Engineers (2019)

[14] Vega, M.R.O., Parise, K., Ramos, L. B., Boff, U., Mattedi, S., Schaeffer, L.,

Malfatti, C.F.: Protic ionic liquids used as metal-forming green lubricants for aluminum: Effect of anion chain length. Mater. Res. 20, 675–687 (2017). doi: 10.1590/1980-5373-MR-2016-0626

[15] Guo, H., Smith, T.W., Iglesias, P.: The study of hexanoate-based protic ionic liquids used as lubricants in steel-steel contact. J. Mol. Liq. 299, 112208 (2019). doi:10.1016/j.molliq.2019.112208

[16] Rocha, M.A.A., Neves, C.M.S.S., Freire, M.G., Russina, O., Triolo, A., Coutinho, J.A.P., Santos, L.M.N.B.F.: Alkylimidazolium based ionic liquids: Impact of cation symmetry on their nanoscale structural organization. J. Phys. Chem. B. 117, 10889–10897 (2013). doi:10.1021/jp406374a

[17] Barnhill, W.C., Qu, J., Luo, H., Meyer, H.M., Ma, C., Chi, M., Papke, B. L.: Phosphonium-organophosphate ionic liquids as lubricant additives: Effects of cation structure on physicochemical and tribological characteristics. ACS Appl. Mater. Interfaces. 6, 22585–22593 (2014). doi: 10.1021/am506702u

[18] Barnhill, W.C., Luo, H., Meyer, H. M., Ma, C., Chi, M., Papke, B.L., Qu, J.: Tertiary and Quaternary Ammonium-Phosphate Ionic Liquids as Lubricant Additives. Tribol. Lett. 63, 1–11 (2016). doi:10.1007/s11249-016-0707-6

[19] Xue, L., Gurung, E., Tamas, G., Koh, Y.P., Shadeck, M., Simon, S.L., Maroncelli, M., Quitevis, E.L.: Effect of Alkyl Chain Branching on Physicochemical Properties of Imidazolium-Based Ionic Liquids. J. Chem. Eng. Data. 61, 1078–1091 (2016). doi:10.1021/acs.jced.5b00658

[20] Xia, Y., Wang, Z., Song, Y.: Influence of hydroxyl group functionalization and alkyl chain length on physicochemical and antiwear properties of hexafluorophosphate imidazolium ionic liquids. Ind. Lubr. Tribol. 66, 443–451 (2014). doi:10.1108/ilt-11-2011-0089

[21] De La Hoz, A.T., Brauer, U.G., Miller, K.M.: Physicochemical and thermal properties for a series of 1-alkyl-4-methyl-1, 2,4-triazolium bis (trifluoromethylsulfonyl)imide ionic liquids. J. Phys. Chem. B. 118, 9944–9951 (2014). doi:10.1021/jp505592t

[22] Er, H., Xu, Y., Zhao, H.: Properties of mono-protic ionic liquids composed of hexylammonium and hexylethylenediaminium cations with trifluoroacetate and bis (trifluoromethylsulfonyl) imide anions. J. Mol. Liq. 276, 379–384 (2019). doi: 10.1016/j.molliq.2018.11.132

[23] Fadeeva, Y.A., Gruzdev, M.S., Kudryakova, N.O., Shmukler, L.E., Safonova, L.P.: Physico-chemical characterization of alkyl-imidazolium protic ionic liquids. J. Mol. Liq. 297, 111305 (2020). doi:10.1016/j.molliq.2019.111305

[24] Xiang, J., Chen, R., Wu, F., Li, L., Chen, S., Zou, Q.: Physicochemical properties of new amide-based protic ionic liquids and their use as materials for anhydrous proton conductors. Electrochim. Acta. 56, 7503–7509 (2011). doi:10.1016/j.electacta.2011.06.103

[25] Xue, Z., Qin, L., Jiang, J., Mu, T., Gao, G.: Thermal, electrochemical and radiolytic stabilities of ionic liquids. Phys. Chem. Chem. Phys. 20, 8382–8402 (2018). doi:10.1039/c7cp07483b

[26] Ma, R., Zhao, Q., Zhang, E., Zheng, D., Li, W., Wang, X.: Synthesis and evaluation of oil-soluble ionic liquids as multifunctional lubricant additives. Tribol. Int. 151, 106446 (2020). doi: 10.1016/j.triboint.2020.106446

[27] Kreivaitis, R., Gumbytė, M., Kupčinskas, A., Kazancev, K., Makarevičienė, V.: Investigating the

tribological properties of PILs derived from different ammonium cations and long chain carboxylic acid anion. Tribol. Int. 141, 105905 (2020). doi:10.1016/j.triboint.2019.105905

[28] Dong, R., Yu, Q., Bai, Y., Wu, Y., Ma, Z., Zhang, J., Zhang, C., Yu, B., Zhou, F., Liu, W., Cai, M.: Towards superior lubricity and anticorrosion performances of proton-type ionic liquids additives for water-based lubricating fluids. Chem. Eng. J. 383, (2020). doi:10.1016/j.cej.2019.123201

[29] Wang, Y., Yu, Q., Cai, M., Shi, L., Zhou, F., Liu, W.: Ibuprofen-Based Ionic Liquids as Additives for Enhancing the Lubricity and Antiwear of Water–Ethylene Glycol Liquid. Tribol. Lett. 65, 1–13 (2017). doi:10.1007/s11249-017-0840-x

[30] Huang, G., Fan, S., Ba, Z., Cai, M., Qiao, D.: Insight into the lubricating mechanism for alkylimidazolium phosphate ionic liquids with different alkyl chain length. Tribol. Int. 140, 105886 (2019). doi:10.1016/j.triboint.2019.105886

[31] Su, T., Song, G., Zheng, D., Ju, C., Zhao, Q.: Facile synthesis of protic ionic liquids hybrid for improving antiwear and anticorrosion properties of water-glycol. Tribol. Int. 153, 106660 (2021). doi:10.1016/j.triboint.2020.106660

[32] Ortega Vega, M.R., Mattedi, S., Schroeder, R.M., de Fraga Malfatti, C.: 2-Hydroxyethilammonium Oleate Protic Ionic Liquid As Corrosion Inhibitor for Aluminum in Neutral Medium. Mater. Corros. (2020). doi:10.1002/maco.202011847

[33] Viesca, J.L., Oulego, P., González, R., Guo, H., Battez, A.H., Iglesias, P.: Miscibility, corrosion and environmental properties of six hexanoate- and sulfonate-based protic ionic liquids. J. Mol. Liq. 114561 (2020). doi:10.1016/j.molliq.2020.114561

[34] Ghanem, O. Ben, Mutalib, M.I.A., Lévêque, J.-M., El-Harbawi, M.: Development of QSAR model to predict the ecotoxicity of Vibrio fischeri using COSMO-RS descriptors. Chemosphere. 170, 242–250 (2017). doi:10.1016/j.chemosphere.2016.12.003

[35] Hodyna, D., Kovalishyn, V., Semenyuta, I., Blagodatnyi, V., Rogalsky, S., Metelytsia, L.: Imidazolium ionic liquids as effective antiseptics and disinfectants against drug resistant *S. aureus*: In silico and in vitro studies. Comput. Biol. Chem. 73, 127–138 (2018). doi:10.1016/j.compbiolchem.2018.01.012

[36] Cao, L., Zhu, P., Zhao, Y., Zhao, J.: Using machine learning and quantum chemistry descriptors to predict the toxicity of ionic liquids. J. Hazard. Mater. 352, 17–26 (2018). doi:10.1016/j.jhazmat.2018.03.025

[37] Abramenko, N., Kustov, L., Metelytsia, L., Kovalishyn, V., Tetko, I., Peijnenburg, W.: A review of recent advances towards the development of QSAR models for toxicity assessment of ionic liquids. J. Hazard. Mater. 384, 121429 (2020). doi:10.1016/j.jhazmat.2019.121429

[38] Mena, I.F., Diaz, E., Palomar, J., Rodriguez, J.J., Mohedano, A.F.: Cation and anion effect on the biodegradability and toxicity of imidazolium– and choline–based ionic liquids. Chemosphere. 240, 124947 (2020). doi:10.1016/j.chemosphere.2019.124947

[39] Tzani, A., Elmaloglou, M., Kyriazis, C., Aravopoulou, D., Kleidas, I., Papadopoulos, A., Ioannou, E., Kyritsis, A., Voutsas, E., Detsi, A.: Synthesis and structure-properties relationship studies of biodegradable hydroxylammonium-based protic ionic liquids. J. Mol. Liq. 224, 366–376 (2016). doi:10.1016/j.molliq.2016.09.086

[40] Avilés, M.D., Carrión-Vilches, F.J., Sanes, J., Bermúdez, M.D.: Diprotic

Ammonium Succinate Ionic Liquid in Thin Film Aqueous Lubrication and in Graphene Nanolubricant. Tribol. Lett. 67, 1–10 (2019). doi:10.1007/s11249-019-1138-y

[41] Khan, A., Gusain, R., Sahai, M., Khatri, O.P.: Fatty acids-derived protic ionic liquids as lubricant additive to synthetic lube base oil for enhancement of tribological properties. J. Mol. Liq. 293, 111444 (2019). doi:10.1016/j.molliq.2019.111444

[42] Ortega Vega, M.R., Ercolani, J., Mattedi, S., Aguzzoli, C., Ferreira, C.A., Rocha, A.S., Malfatti, C.F.: Oleate-Based Protic Ionic Liquids As Lubricants for Aluminum 1100. Ind. Eng. Chem. Res. 57, 12386–12396 (2018). doi:10.1021/acs.iecr.8b02426

[43] Yao, Y., Xu, Y., Fan, X., Zhu, M., Liu, G.: Tribological properties of spherical and mesoporous NiAl particles as ionic liquid additives. Friction. 8, 384–395 (2020). doi:10.1007/s40544-019-0266-6

[44] Viesca, J.L., Battez, A.H., González, R., Reddyhoff, T., Pérez, A.T., Spikes, H.A.: Assessing boundary film formation of lubricant additivised with 1-hexyl-3-methylimidazolium tetrafluoroborate using ECR as qualitative indicator. Wear. 269, 112–117 (2010). doi:10.1016/j.wear.2010.03.014

[45] Zhou, Y., Leonard, D.N., Guo, W., Qu, J.: Understanding Tribofilm Formation Mechanisms in Ionic Liquid Lubrication. Sci. Rep. 7, 1–8 (2017). doi:10.1038/s41598-017-09029-z

[46] Guo, W., Zhou, Y., Sang, X., Leonard, D.N., Qu, J., Poplawsky, J.D.: Atom Probe Tomography Unveils Formation Mechanisms of Wear-Protective Tribofilms by ZDDP, Ionic Liquid, and Their Combination. ACS Appl. Mater. Interfaces. 9, 23152–23163 (2017). doi:10.1021/acsami.7b04719

[47] Yu, Q., Wang, Y., Huang, G., Ma, Z., Shi, Y., Cai, M., Zhou, F., Liu, W.: Task-Specific Oil-Miscible Ionic Liquids Lubricate Steel/Light Metal Alloy: A Tribochemistry Study. Adv. Mater. Interfaces. 5, 1–12 (2018). doi:10.1002/admi.201800791

[48] Kawada, S., Watanabe, S., Kondo, Y., Tsuboi, R., Sasaki, S.: Tribochemical Reactions of Ionic Liquids Under Vacuum Conditions. Tribol. Lett. 54, 309–315 (2014). doi:10.1007/s11249-014-0342-z

[49] Kawada, S., Watanabe, S., Tadokoro, C., Tsuboi, R., Sasaki, S.: Lubricating mechanism of cyano-based ionic liquids on nascent steel surface. Tribol. Int. 119, 474–480 (2018). doi:10.1016/j.triboint.2017.11.019

[50] Sharma, V., Doerr, N., Aswath, P.B.: Chemical–mechanical properties of tribofilms and their relationship to ionic liquid chemistry. RSC Adv. 6, 22341–22356 (2016). doi:10.1039/C6RA01915C

[51] Landauer, A.K., Barnhill, W.C., Qu, J.: Correlating mechanical properties and anti-wear performance of tribofilms formed by ionic liquids, ZDDP and their combinations. Wear. 354–355, 78–82 (2016). doi:10.1016/j.wear.2016.03.003

[52] Huang, G., Yu, Q., Ma, Z., Cai, M., liu, W.: Probing the lubricating mechanism of oil-soluble ionic liquids additives. Tribol. Int. 107, 152–162 (2017). doi:10.1016/j.triboint.2016.08.027

[53] Li, W., Kumara, C., Meyer, H.M., Luo, H., Qu, J.: Compatibility between Various Ionic Liquids and an Organic Friction Modifier as Lubricant Additives. Langmuir. 34, 10711–10720 (2018). doi:10.1021/acs.langmuir.8b02482

[54] Guo, H., Liu, R., Fuentes-Aznar, A., Iglesias Victoria, P.: Friction and Wear Properties of Halogen-Free and

Halogen-Containing Ionic Liquids Used As Neat Lubricants, Lubricant Additives and Thin Lubricant Layers. In: Volume 10: 2017 ASME International Power Transmission and Gearing Conference. pp. 1–5. American Society of Mechanical Engineers (2017)

[55] Duan, H., Li, W., Kumara, C., Jin, Y., Meyer, H.M., Luo, H., Qu, J.: Ionic liquids as oil additives for lubricating oxygen-diffusion case-hardened titanium. Tribol. Int. 136, 342–348 (2019). doi:10.1016/j.triboint.2019.03.069

[56] González, R., Viesca, J.L., Battez, A. H., Hadfield, M., Fernández-González, A., Bartolomé, M.: Two phosphonium cation-based ionic liquids as lubricant additive to a polyalphaolefin base oil. J. Mol. Liq. 293, 26–30 (2019). doi: 10.1016/j.molliq.2019.111536

[57] Guo, H., Adukure, A.R., Iglesias, P.: Effect of Ionicity of Three Protic Ionic Liquids as Neat Lubricants and Lubricant Additives to a Biolubricant. Coatings. 9, 713 (2019). doi:10.3390/coatings9110713

[58] Viesca, J.L., Mallada, M.T., Blanco, D., Fernández-González, A., Espina-Casado, J., González, R., Hernández Battez, A.: Lubrication performance of an ammonium cation-based ionic liquid used as an additive in a polar oil. Tribol. Int. 116, 422–430 (2017). doi:10.1016/j.triboint.2017.08.004

[59] Magar, S.A., Guo, H., Iglesias, P.: Ionic liquid as cutting fluid additive using minimum quantity lubricant (MQL) in titanium-ceramic contact. ASME Int. Mech. Eng. Congr. Expo. Proc. 12, 1–7 (2019). doi:10.1115/IMECE2019-10647

[60] Sernaglia, M., Blanco, D., Hernández Battez, A., González, R., Fernández-González, A., Bartolomé, M.: Two fatty acid anion-based ionic liquids - part II: Effectiveness as an additive to a polyol ester. J. Mol. Liq. 310, (2020). doi:10.1016/j.molliq.2020.113158

[61] Zhu, L., Zhao, G., Wang, X.: Investigation on three oil-miscible ionic liquids as antiwear additives for polyol esters at elevated temperature. Tribol. Int. 109, 336–345 (2017). doi:10.1016/j.triboint.2016.10.032

[62] Amiril, S.A.S., Rahim, E.A., Embong, Z., Syahrullail, S.: Tribological investigations on the application of oil-miscible ionic liquids additives in modified Jatropha-based metalworking fluid. Tribol. Int. 120, 520–534 (2018). doi:10.1016/j.triboint.2018.01.030

[63] Hansen, J., Björling, M., Minami, I., Larsson, R.: Performance and mechanisms of silicate tribofilm in heavily loaded rolling/sliding non-conformal contacts. Tribol. Int. 123, 130–141 (2018). doi:10.1016/j.triboint.2018.03.006

[64] Cigno, E., Magagnoli, C., Pierce, M. S., Iglesias, P.: Lubricating ability of two phosphonium-based ionic liquids as additives of a bio-oil for use in wind turbines gearboxes. Wear. 376–377, 756–765 (2017). doi:10.1016/j.wear.2017.01.010

[65] Khan, A., Yasa, S.R., Gusain, R., Khatri, O.P.: Oil-miscible, halogen-free, and surface-active lauryl sulphate-derived ionic liquids for enhancement of tribological properties. J. Mol. Liq. 318, 114005 (2020). doi:10.1016/j.molliq.2020.114005

[66] Zheng, G., Zhang, G., Ding, T., Xiang, X., Li, F., Ren, T., Liu, S., Zheng, L.: Tribological properties and surface interaction of novel water-soluble ionic liquid in water-glycol. Tribol. Int. 116, 440–448 (2017). doi:10.1016/j.triboint.2017.08.001

[67] Fan, M., Du, X., Ma, L., Wen, P., Zhang, S., Dong, R., Sun, W., Yang, D., Zhou, F., Liu, W.: In situ preparation of

multifunctional additives in water. Tribol. Int. 130, 317–323 (2019). doi: 10.1016/j.triboint.2018.09.020

[68] Wang, Y., Yu, Q., Cai, M., Zhou, F., Liu, W.: Halide-free PN ionic liquids surfactants as additives for enhancing tribological performance of water-based liquid. Tribol. Int. 128, 190–196 (2018). doi:10.1016/j.triboint.2018.07.018

[69] Wang, Y., Yu, Q., Ma, Z., Huang, G., Cai, M., Zhou, F., Liu, W.: Significant enhancement of anti-friction capability of cationic surfactant by phosphonate functionality as additive in water. Tribol. Int. 112, 86–93 (2017). doi:10.1016/j.triboint.2017.03.034

[70] Avilés, M.D., Carrión, F.J., Sanes, J., Bermúdez, M.D.: Effects of protic ionic liquid crystal additives on the water-lubricated sliding wear and friction of sapphire against stainless steel. Wear. 408–409, 56–64 (2018). doi:10.1016/j.wear.2018.04.015

[71] Zheng, D., Wang, X., Zhang, M., Ju, C.: Synergistic Effects Between the Two Choline-Based Ionic Liquids as Lubricant Additives in Glycerol Aqueous Solution. Tribol. Lett. 67, 1–13 (2019). doi:10.1007/s11249-019-1161-z

[72] Kreivaitis, R., Gumbytė, M., Kupčinskas, A., Kazancev, K., Ta, T.N., Horng, J.H.: Investigation of tribological properties of two protic ionic liquids as additives in water for steel–steel and alumina–steel contacts. Wear. 456–457, (2020). doi:10.1016/j.wear.2020.203390

[73] Zheng, D., Wang, X., Liu, Z., Ju, C., Xu, Z., Xu, J., Yang, C.: Synergy between two protic ionic liquids for improving the antiwear property of glycerol aqueous solution. Tribol. Int. 141, 105731 (2020). doi:10.1016/j.triboint.2019.04.015

[74] Qu, J., Bansal, D.G., Yu, B., Howe, J. Y., Luo, H., Dai, S., Li, H., Blau, P.J., Bunting, B.G., Mordukhovich, G., Smolenski, D.J.: Antiwear performance and mechanism of an oil-miscible ionic liquid as a lubricant additive. ACS Appl. Mater. Interfaces. 4, 997–1002 (2012). doi:10.1021/am201646k

[75] Yu, B., Bansal, D.G., Qu, J., Sun, X., Luo, H., Dai, S., Blau, P.J., Bunting, B. G., Mordukhovich, G., Smolenski, D.J.: Oil-miscible and non-corrosive phosphonium-based ionic liquids as candidate lubricant additives. Wear. 289, 58–64 (2012). doi:10.1016/j.wear.2012.04.015

[76] Cowie, S., Cooper, P.K., Atkin, R., Li, H.: Nanotribology of Ionic Liquids as Lubricant Additives for Alumina Surfaces. J. Phys. Chem. C. 121, 28348–28353 (2017). doi:10.1021/acs.jpcc.7b09879

[77] Zhang, Y., Cai, T., Shang, W., Sun, L., Liu, D., Tong, D., Liu, S.: Environmental friendly polyisobutylene-based ionic liquid containing chelated orthoborate as lubricant additive: Synthesis, tribological properties and synergistic interactions with ZDDP in hydrocarbon oils. Tribol. Int. 115, 297–306 (2017). doi:10.1016/j.triboint.2017.05.038

[78] Qu, J., Barnhill, W.C., Luo, H., Meyer, H.M., Leonard, D.N., Landauer, A.K., Kheireddin, B., Gao, H., Papke, B. L., Dai, S.: Synergistic Effects between Phosphonium-Alkylphosphate Ionic Liquids and Zinc Dialkyldithiophosphate (ZDDP) as Lubricant Additives. Adv. Mater. 27, 4767–4774 (2015). doi: 10.1002/adma.201502037

[79] Taher, M., Shah, F.U., Filippov, A., De Baets, P., Glavatskih, S., Antzutkin, O.N.: Halogen-free pyrrolidinium bis (mandelato)borate ionic liquids: Some physicochemical properties and lubrication performance as additives to polyethylene glycol. RSC Adv. 4, 30617–30623 (2014). doi:10.1039/c4ra02551b

[80] Guo, H., Pang, J., Adukure, A.R., Iglesias, P.: Influence of Hydrogen

Bonding and Ionicity of Protic Ionic Liquids on Lubricating Steel–Steel and Steel–Aluminum Contacts: Potential Ecofriendly Lubricants and Additives. Tribol. Lett. 68, 1–10 (2020). doi: 10.1007/s11249-020-01354-1

[81] Wang, Y., Yu, Q., Ma, Z., Huang, G., Cai, M.: Tribology International Signi fi cant enhancement of anti-friction capability of cationic surfactant by phosphonate functionality as additive in water. Tribiology Int. 112, 86–93 (2017). doi:10.1016/j.triboint.2017.03.034

[82] Espinosa, T., Jimenez, M., Sanes, J., Jimenez, A.E., Iglesias, M., Bermudez, M.D.: Ultra-low friction with a protic ionic liquid boundary film at the water-lubricated sapphire-stainless steel interface. Tribol. Lett. 53, 1–9 (2014). doi:10.1007/s11249-013-0238-3

[83] Guo, H., Chen, F., Liu, R., Iglesias, P.: Lubricating Ability of Magnesium Silicate Hydroxide–Based Nanopowder as Lubricant Additive in Steel–Steel and Ceramic–Steel Contacts. Tribol. Trans. 63, 585–596 (2020). doi:10.1080/10402004.2019.1710312

[84] Taaber, T., Enok, A.E., Joost, U., Oras, S., Järvekülg, M., Lõhmus, R., Mäeorg, U., Saal, K.: Tribological properties of protic ionic liquid and functionalized copper oxide nanoparticles as additives to base oil. Mechanika. 21, 148–153 (2015). doi: 10.5755/j01.mech.21.2.11723

[85] Han, X., Zhang, Z., Thrush, S.J., Barber, G.C., Qu, H.: Ionic liquid stabilized nanoparticle additive in a steel-ceramic contact for extreme pressure application. Wear. 452–453, 203264 (2020). doi:10.1016/j.wear.2020.203264

Chapter 5

Applications of Ionic Liquids in Gas Chromatography

Umaima Gazal

Abstract

The environment offers an enormous innovative panorama of prospects intended for the research of novel biodegradable diluents. Regular composites have been lately recycled to formulate the anionic and cationic fraction of RTIL. Numerous applications of ionic liquids have been explored in segregation discipline. Attributable to the extraordinary polarization as well as exceptional current steadiness, IL-centered immobile segments have been applied to resolution of varied series of critically stimulating complexes frequently extremely polar composites using great boiling points plus physical resemblances comprising elongated sequence fatty acids, essential oils, polycyclic aromatic sulfur heterocycles (PASHs) and PCBs. IL-centered immobile segments facilitated the gas chromatography study for effective as well as precise amount of liquid in the industrialized yields for example pharmaceutical as well as petrochemicals complexes.

Keywords: ionic liquids, gas chromatography, static stages, Zwitterionic liquids, polymeric ionic liquids

1. Introduction

Ionic liquids are the utmost promising liquid green solvents with wide applications in separation science. The effect of the IL organic configuration as well as the stimulus of tributary factors, for example the IL temperature, pH, concentration, analysis time and voltage, are compatibly rationally talked concerning the accomplished parting enlargements. Gas chromatography is unique and extreme proficient, dependable, as well as stout methods for the study of unstable plus semi-volatile composites. Effective along with rapid gas chromatographic investigation of objective analytes is mostly reliant on the enactment of the gas chromatography column. Though here have be present main active developments, there quiet a solid claim of extremely choosy, indolent, polar also thermally constant gas chromatography pillars intended for critically stimulating composites for example polychlorinated biphenyls, unrestricted fatty acids and unstable amines [1]. Furthermore, the physicochemical characteristics for instance surface tension, viscosity and melting point are too acute to yield extremely proficient gas chromatography columns. The viscosities of furthermost ionic liquids are frequently 1–3 remits of scale greater than outmoded biological diluents [2]. In demand for an ionic liquid to be measured as a immobile stage, the solid must have great viscosity that rests fixed above a comprehensive high temperature choice. Van der Waals as well as Hydrogen bonding kind interfaces amongst the anion plus cation of Ionic liquids rule the viscosity-properties. Furthermore, it is imperative to ruminate the surface tension

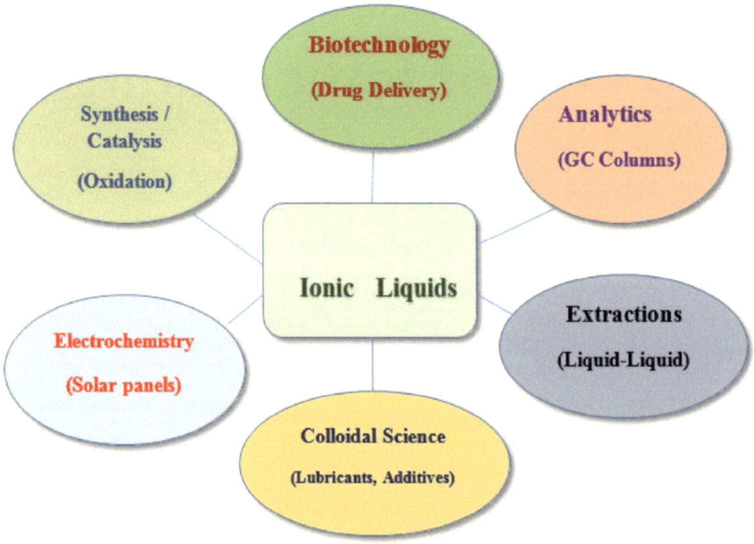

Figure 1.
Applications of ionic liquids in various fields.

of the Ionic liquids. Its values extending as of 30 to 50 dyne/cm usually display bigger wettability on the barrier of unprocessed tube pillars [2]. Ionic liquids establish an assembly of biological salts which are fluid lower than 100 °C, moreover, the ionic liquids that are fluid at room temperature are generally recognized as room temperature ionic liquids [3]. Ionic liquids are easy to manufacture, thermally steady, flameproof, chemically inactive, retain small vapor density, polar, and their discernment can be simply regulated by means of fluctuating the component anion or cation; and from now they have been extensively recycled as static stages in conservative gas chromatography (**Figure 1**) [4–10].

Ionic liquids can also be recycled as diluents for the suspension of various resources for example fiber [11], chitin [12], etc. The outstanding solubility of biological/inert composites in ionic liquids as well as a extensive variety of it as the fluid state brand them noble diluents for several responses. Furthermore, they displayed modest produces when related with conservative biological diluents [13, 14]. Thermodynamic factors of these ionic liquids were studied through chromatographic methods. ILs are centered on the numerous method, in company with the utmost extensively considered are N-alkylpyridinium, alkylammonium, N'N-dialkylimidazolium, and alkylphosphonium.

2. Stationary phase in gas chromatography using ILs

Owing to the ever-developing mandate for the great determination, high sympathy, as well as statistics amusing investigation of composite models for instance aromas, smells, petrochemicals, plus pharmacological uncooked supplies, continuous expansions of gas chromatography supports through exclusive discrimination, squat bleed, high dullness, and in addition varied high temperature operational series are desirable. Because of the high polarization then exceptional thermal constancy, IL-centered stationary stages have been employed to decide an extensive series of methodically stimulating complexes typically precise polar complexes with high steaming facts also fundamental resemblances comprising lengthy sequence oily acids, vital oils and polycyclic aromatic sulfur heterocycles [15, 16]. ILs are characteristically organized with a phosphorous- or nitrogen-comprising biological cation as well as an inorganic or else organic anion. Meanwhile the chief outline of IL-centered GC supports in 1999, ILs have been effectively engaged as stationary stages because of their trivial high polarity, tunable selectivity, high thermal stability and vapor pressure [17]. The ILs can be altered through diverse efficient collections to endure countless solvation interfaces in addition to display exclusive chromatographic discernment such as GC immobile phases. ILs have been exposed to have exclusive solvation competences plus discernment's on the distinctive solvent/solute interfaces [18]. Imidazolium-centered ILs be able to as per contrived to discrete equally non-polar and polar analytes [17]. As, by means of varying the anionic lot of the imidazolium IL since chloride [Cl]- to hexafluorophosphate [PF_6]-, a substantial variance in discernment was perceived for polar analytes equaled to the non-polar complexes (**Figure 2**).

Additional examination of dissimilar modules of ILs, containing monocationic imidazolium, pyridinium, as well as pyrrolidinium exposed that the hydrogen contributor capability of the IL immobile phases was subjugated through IL cation. In contrast, the anionic lot was create to adopt the part of hydrogen acceptor anion as of proton giver analytes for example carboxylic acids plus alcohols [2]. Consequently, dicationic [19], tricationic [20], as well as phosphonium-centered cations [21] were oppressed to expand great thermal constancy plus fluid variety of ILs equated to customary monocationic static levels. Lately, in an effort to extend the applicability of IL static phases, task-specific ionic liquids (TSILs) were familiarized by functionalizing the IL cation with numerous agents [22]. For instance, the integration of aromatic segments in the IL cation improved the discernment for scented complexes, for example polycyclic aromatic hydrocarbons (PAHs). This is owing to enriched π-π sort communications amongst analytes as well as the aromatic clusters of the IL cation [4]. Overview of polar efficient clusters, for example hydroxyl segments, can effect in enlarged discernment for hydrogen compliant analytes [23]. Consequently, tweaking the IL-centered GC static phase configuration might augment choosiness essential for parting of precise compound model elements with comparable polarizations.

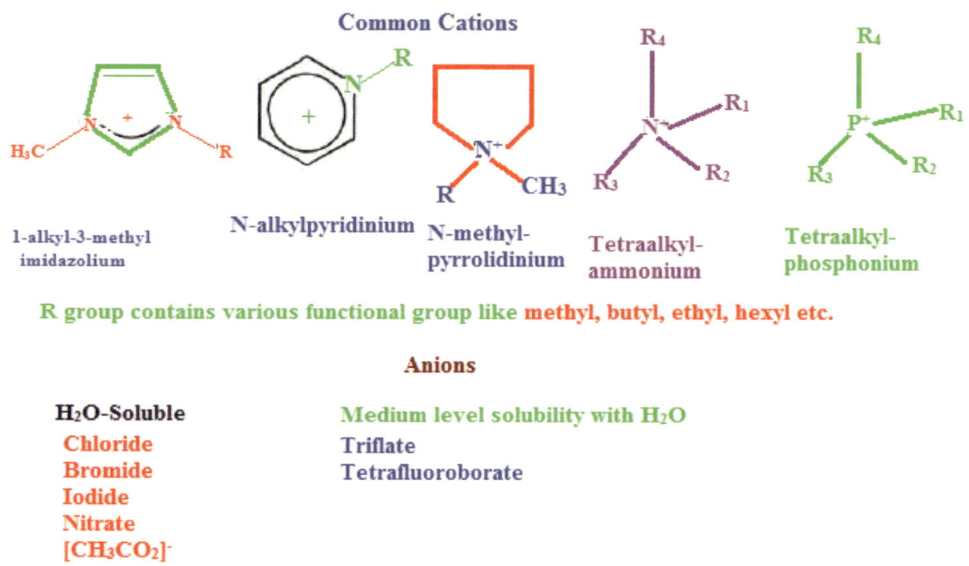

Figure 2.
Common cations and anions.

One of the important physical property of static stage is melting point as it basically commands the least effective high temperature of the ensued GC column. ILs with short melting points are extremely required as well as are usually acquired via integrating proportion-flouting sections also alkyl sideways chains with diverse dimensions [24–26]. Analytes naturally intermingle using IL-centered stationary segments over moreover partition- or adsorption-kind contrivance [6, 27–28]. Better parting efficacies existed usually providing through the partition-kind retaining contrivance. Once the furnace temperature is lesser than the melting point of the IL-centered static stage, the molecular interface amongst the analytes as well as static stage is to be expected to be controlled by means of adsorption. Variance perusing calorimetry is usually operated to define the melting point of IL-centered stationary stages [25].

3. Incorporation of ionic liquids in multidimensional gas chromatography

Multidimensional gas chromatography is an influential method to accomplish progressive parting of impulsive as well as quasi-impulsive composites in compound environments [29–31]. By means of Multidimensional gas chromatography method, two or else extra gas chromatographic partings are engaged in a consecutive manner [29]. The paramount requirement to effectively enhance peak capacity in the composite system is to employ a combination of GC stationary phases possessing different selectivities. It was presented to the chromatographic state compromises advanced ultimate ability than conservative one-dimensional gas chromatography, permitting on behalf of the determination of model elements by means of comparable polarizations otherwise instabilities [32]. In this method, analytes are evaporated then exposed to a sequence of gas chromatography supports by means of chemically diverse static stages attached over an edge. In multidimensional gas chromatography analyte parting is preserved on both column, prominent to an upsurge in the parting control associated to the one [33]. Two types of multidimensional gas chromatography are usually engaged, specifically, core-wounding also inclusive. In core-wounding multidimensional gas chromatography, merely a choice rare portions of overflow after the first support are transported to the second support for extra parting [34].

Diverse support selectivity, which can be characterized using a liberated parting procedure, is the significant requisite to acquire advanced top capability in multidimensional gas chromatography methods [35]. Several customary non-ionic gas chromatography static segments are categorized as both non-polar and polar segments. These supports display a deficiency of variety in positions of solvation abilities, which can bound their capacity to decide composite models through gas chromatography× gas chromatography. Because of this hitch, ionic liquids centered supports have appeared as alternate gas chromatography × gas chromatography static stages. Utilizing ionic liquids centered stakes can permit exclusive solvation abilities in addition selectivities, moreover to advance thermal constancies comparative to customary segments. Ionic liquids have been applied as existing segments combined with customary non-ionic segments in numerous gas chromatography× gas chromatography partings [36]. Meanwhile maximum gas chromatography × gas chromatography partings can be controlled constructed on analyte instability in the first measurement monitored through involvement of dissimilar relations in the second measurement, it is collective to practice ionic liquids supports as the second measurement to estimate their enactment in relations of retaining contrivances. Presently, a huge amount of profitably accessible ionic liquids supports, for example the Supelco Low Bleed community, comprise numerous phosphonium and imidazolium grounded di cations which are typically combined with frequently tri fluoromethanesulfonate and bis[(trifluoromethyl)sulfonyl]imide anions [2, 20]. Economical ionic liquids static stages have been engaged in the parting of total of analytes for instance savor plus fragrance amalgams [36], aromatic hydrocarbons [37], alkyl halides [38], alkyl phosphonates [39], fatty acid methyl esters [40], as well as additional polar analytes (nitrogen, sulfur as well as oxygen-comprising composites). These analysis specify that ionic liquids supports establish considerable advanced selectivity plus retaining in the direction of commonly polar analytes associated to non-polar analytes owing to hydrogen-bonding interaction, electrostatic interactions and dipole–dipole relations, amongst ions [41].

4. Polymeric ionic liquids centered static stages in gas chromatography

Polymeric ionic liquids are stimulating family of composites that can be recycled as sorbent coverings in solid phase micro extraction. Polymeric ionic liquids are artificial polymers manufactured after ionic liquids monomers. Furthermore, Polymeric ionic liquids can be basically modified to display greater sensitivity and selectivity nearby diverse section of analytes. Polymeric ionic liquids are characteristically manufactured through functionalizing a polymerizable practical cluster on the cationic component of the ionic liquid by free radical polymerization in the attendance of a thermal originator. Polymeric ionic liquids reveal greater thermal constancy in addition to a confrontation to viscosity decrease at greater temperatures. These valuable structures can develop fiber lifespan, toughness plus eligibility of Polymeric ionic liquids while retentive the discrimination relics the intrinsic to ionic liquids. Polymeric ionic liquids have been displayed to extant extraordinary possessions as well as exhibitions [42–48], assisting novel plus stimulating parting procedures [49–51]. Ionic liquids have drawn abundant consideration in latest centuries as constituents for static stages in gas chromatography, because of stuffs similar their capability to create concurrent nonpolar as well as polar interfaces with the analytes, their extraordinary thermal constancy, before their insignificant air compressions then extensive fluid series [52–56]. Also, it is price revealing that these things can be effortlessly well-adjusted over minor fluctuations in the assembly of either the anion or cation, which, also, can intensely modify the choosiness or else the parting capability for the analyte of concern [36, 57–59]. The concern in ionic liquid-covered gas chromatography supports has enlarged afterward their marketable outline in 2008, also today, numerous ionic liquids glazed supports with dissimilar features are viably accessible. Though, one main task for the growth of static stages built on ionic liquids is the research of extremely standardized coverings, which would service decent ultimate regularities as well as extremely active complex partings, also concurrently, deliver extraordinary thermal constancies for the subsequent gas chromatography supports [60–62]. At extraordinary temperatures, identical ionic liquid-glazed silica supports can practice flick commotion prominent to a diminution in the analyte retaining periods as well as efficacy. In this respect, polymerized ionic liquids can offer the compulsory replies, preserving the outstanding thermal constancy of the supports, in addition to uniting the chief structures of an ionic liquid as well as the distinctive polymer characteristics for example better automated constancy plus development capability [62–65].

5. Preference of Zwitterionic liquids in gas chromatography

Zwitterions, consequent after ionic liquids have inimitable characteristics for example reasonably small crystal conversion temperature, slight ion conductivity as well as exclusive stage conduct afterward partying with water. Moreover, the combination of convinced zwitterions plus negligible quantity of water can be observed as an exceptional liquid ideal of cell tissues. Zwitterionic liquids can be chemically precise like to predictable aprotic ionic liquids, excluding that the negative as well as positive charges exist in on the similar particle. Zwitterionic complexes are elements that have an entire clear charge of zero as well as are therefore electrically impartial. They transmit proper electrical charges of reverse symbol contained on diverse particles as well as formerly can be measured as internal salts. The utmost collective zwitterionic-kind ionic liquids are nitrogene heterocycles with sulfonate component. These Zwitter ionic Liquids have been competently considered as designable electrolyte constituents for fuel cells [66] as well as lithium batteries [67]. Not as much of discovered Zwitter ionic Liquids are configurations founded on imidazoliums with carboxylate occupations. They have been cast-off as forde sulfurization of fuels [68], Bronsted acidic catalytic agent [69], as liquid crystals [70] or for metallic oxides solubilization [71]. Therefore amino acids be existent typically as per zwitterions in a definite variety of pH then the pH at which the regular charge is zero is known as the particle's isoelectric fact (**Figure 3**).

Through zwitterionic composites, anions as well as cations are roped covalently. Imidazolium sulfonate is one of the example of zwitterionic liquids. The production of room temperature zwitterionic liquids in which both sulfonate anion as well as imidazolium cation elements attribute to the parental particle was conveyed in literature (**Figure 4**) [72, 73].

Motivated via means of the proclamation that these composites can be organized as fluids at room temperature, three zwitterionic liquids integrating alkyl side chain and oligoether substituents were intended then inspected as gas chromatography static stages. The fundamentally-regulated zwitterionic liquids -centered static stages offer distinctive choosiness, robust retaining, exceptional top regularity, also a reasonably widespread employed series appropriate for the study of volatile carboxylic acids. This comprising volatile fatty acids for instance lactic acid as well as butyric acid are significant for the construction of cosmetics, pharmaceuticals and fuels [74–76]. Gas chromatography is furthermost usually recycled for the quantification as well as parting of specific acids in acylated lipids. Derivatization of volatile carboxylic acids by means of numerous approaches for example alkylation plus acylation is characteristically accomplished to upsurge the explosive nature of these composites in addition to mark their investigation viable through gas chromatography (**Table 1**).

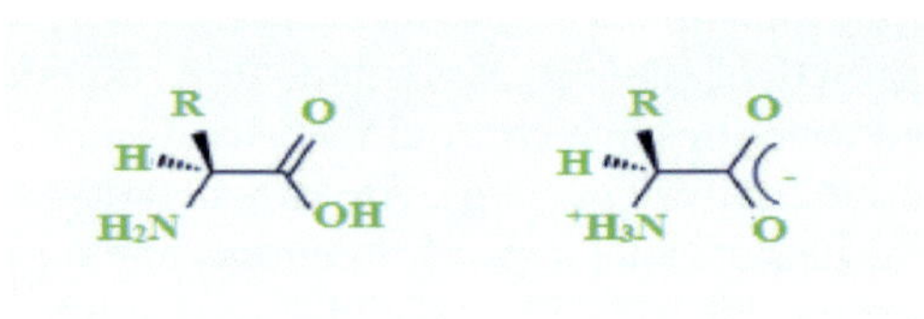

Figure 3.
Zwitter ionic liquids.

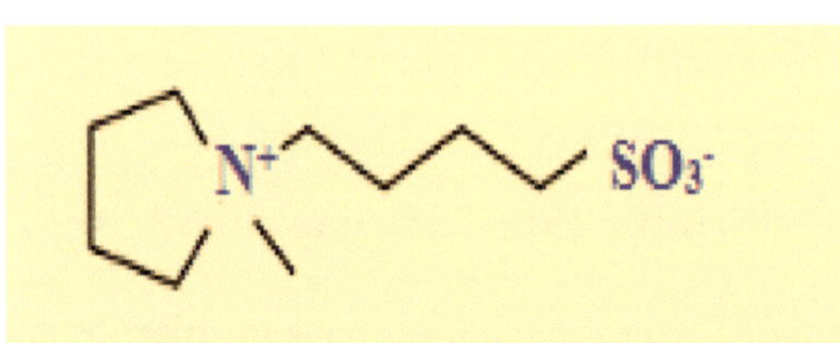

Figure 4.
Imidazolium sulfonate Zwitterionic liquids.

S. No	Embattled Composites	Ionic liquids	Uses	Reference
1.	Methyl esters / Fatty acids	1,9-di(3-vinyl-imidazolium) nonane bis(trifluoromethyl) sulfonylimidate	Gas chromatography stage	[77]
2.	Aromatic Composites / Alcohols	[BMIM][NTf_2], [BMIM]Cl	Gas chromatography stage	[78]

Table 1.
Use of ionic liquids as superficial-integrated static segments.

6. Green aspects of ionic liquids

The adjustable physicochemical properties of ionic liquids have prolonged their usage addicted to a wide variety of diverse uses. Ionic liquids have an abundant prospective in biological amalgamation, electrochemistry, mass spectrometry, green chemistry as well as partings [79, 80]. In the field of analytical chemistry, ionic liquids have been recycled as static segments as well as diluents for headspace gas chromatography [80], movable segment extracts plus external-attached static segments in liquid chromatography [79, 81] for liquid–liquid abstractions as well as solid-phase micro abstraction [79, 80, 82]. Several "green-engrossed" manufacturing have originate that ionic liquids are outstanding applicants for their uses because of their precise small vapor compression [83, 84]. The overview of ionic liquids such as static segments has released up and around novel outlooks in this arena by means of their exclusive solvation features outcome in unusual discernment, which is entirely dissimilar to that of typical polyethylene glycol as well as poly dimethyl siloxane centered supports. Since of their atypical discernment plus extraordinary unresponsiveness, ionic liquids centered supports have previously establish numerous solicitations in the normal item for consumption ground in and multidimensional as well as mono gas chromatography in addition to preparative gas chromatography, prominent to the comprehensive investigation of composite sections (containing aqueous resolutions), plus the parting of stimulating sets of complexes. The speedily growing usage of Ionic liquids equally in educational plus manufacturing arenas have created an increasing apprehension approximately their effect on the environs. Meanwhile Ionic liquids are extremely solvable in liquid however are not continuously ecofriendly, a discharge of ionic liquids into the atmosphere might clue to substantial water contamination complications. Furthermore, Ionic liquids could develop insistent contaminants in discarded water seepage because of their great constancy in water. Intended for this purpose, research inspecting Ionic liquids biodegradability are of inordinate significance. The rising character of Ionic liquids in production as well as study plus the growing alarm nearby their green influence have advised a requisite for the progress of profligate, dependable as well as reproducible techniques for the classification plus investigation of Ionic liquids [85].

7. Conclusion

Ionic liquids have solicitations in various areas in chemistry. The applications of ionic liquids as extracts in chromatography displays abundant rewards equated to further extracts. Ionic liquids have been realistic in diverse extents of parting, for instance ionic liquid sustained tissues, as moveable segment extracts as well as external-joined static stages in chromatography partings also as the abstraction diluent in model provisions, since they can be collected from numerous anions in addition to cations that alter the things as well as stage conduct of fluids. The inflammable, non- explosive environment of ionic liquids marks them an outstanding optimal for the expansion of nontoxic methods. A substantial benefit of ionic liquids-centered stationary segments is their capability to have adequate to high updraft solidity though similarly unveiling a comprehensive host of solvation proficiencies, specific of their inimitable selectivities. In spite of their attainment, viable Ionic liquids-centered immobile segments dearth the solving authority for non-polar analytes, predominantly unsaturated as well as saturated hydrocarbons, cycloalkanes. This nonexistence of discernment has diminished fervor amongst certain parting experts who might modulate the feature of fundamental fine-tuning (in relations of anion/cation combining in addition to operational structures of every constituent) while emerging ionic liquids to display great discernment also robust solving influence. Ionic liquids have strained substantial consideration as gas chromatography immobile segments as of their tunable chemical plus physical properties. Conversely, profitable Ionic liquids-centered gas chromatography supports have not reconnoitered entirely of the solvation characteristics that can be obtainable through ionic liquids.

Moreover, their polarization, viscosity, hydrophobicity plus further physical and chemical properties can be designated by means of selecting the anionic and cationic component. Ionic liquids are considered as "exclusive diluents" as of this adjustable environment, which rises their prospective solicitations. The consumption of ionic liquids is maiden innovative prospects in diverse regions of parting discipline, with novel countless solicitations. Additional uses in partings are linked to the ecological, pharmacological, biomedical as well as various manufacturing trades. Ionic liquids have been discovered in partings for abstraction, reinforced fluid membranes, as extracts then as static segments in chromatography.

Acknowledgements

The author received no financial support for writing this article. I want to acknowledge my mother **Mrs. Nargis Sultana** and husband **Mr. Safdar Hussain** for his encouragement and support for writing this article.

Author details

Umaima Gazal
Department of Chemistry, Raja Bahadur Venkata Rama Reddy Women's College, Affiliated to Osmania University, Hyderabad, India

*Address all correspondence to: dr.umaimagazal@gmail.com

References

[1] Shen B, Semin D, Fang J, Guo G, Analysis of 4-bromo-3-fluorobenzaldehyde and separation of its region isomers by one-dimensional and two-dimensional gas chromatography, J. Chromatogr. A. 2016, 1462: 115-123.

[2] Yao C, Anderson J.L, Retention characteristics of organic compounds on molten salt and ionic liquid-based gas chromatography stationary phases. J. Chromatogr. A. 2009, 1216: 1658-1712.

[3] Hayes R, Warr G.G, Atkin R, Structure and nanostructure in ionic liquids, Chem. Rev. 2015,115: 6357-6426.

[4] Anderson J.L, Armstrong D.W, High-stability ionic liquids. A new class of stationary phases for gas chromatography, Anal. Chem. 2003, 75: 4851-4858.

[5] Anderson J.L, Armstrong D.W, Immobilized ionic liquids as high-selectivity/high temperature/high-stability gas chromatography stationary phases. Anal. Chem. 2005, 77: 6453-6462.

[6] Poole S. K, Shetty P. H, Poole C.F, Chromatographic and spectroscopic studies of the solvent properties of a new series of room-temperature liquid tetraalkylammonium sulfonates, Anal. Chim. Acta. 1989, 218: 241-264.

[7] Pomaville R. M, Poole C. F, Solute-solvent interactions in liquid tetrabutylammonium sulfonate salts studied by gas chromatography. Anal. Chem. 1988, 60: 1103-1108.

[8] Anderson J.L, Ding R, Ellern A, Armstrong D.W, Structure and properties of high stability geminal dicationic ionic liquids, J. Am. Chem. Soc. 2005, 127: 593-604.

[9] Tran C.D, Challa S, Fullerene-impregnated ionic liquid stationary phases for gas chromatography. Analyst. 2008, 133: 455-464.

[10] Breitbach Z.S, Armstrong D.W, Characterization of phosphonium ionic liquids through a linear solvation energy relationship and their use as GLC stationary phases. Anal. Bioanal. Chem. 2008, 390: 1605-1617.

[11] Zhu S, Wu Y, Chen Q, Yu Z, Wang C, Jin S, Ding Y, and Wu G, Utilization of Cellulose by combining two major green chemistry principles: using environmentally preferable solvents and bio-renewable feed-stocks. Green Chemistry, 2006, 8: 325-327.

[12] Xie H, Zhang S, and Li S, Chitin and chitosan dissolved in ionic liquids as reversible sorbents of CO_2. Green Chemistry. 2006, 8: 630-633.

[13] Morrison, D.W, Forbes, D.C, and Davis Jr, J.H, Base-promoted reactions in ionic liquid solvents. The Knoevenagel and Robinson annulation reactions, Tetrahedron Lett. 2001, 42: 6053-6055.

[14] Welton T, Room-Temperature Ionic Liquids. Solvents for Synthesis and Catalysis. Chemical Reviews. 1999, 99: 2071-2084.

[15] T.D. Ho, C. Zhang, L.W. Hantao, J.L. Anderson, Ionic Liquids in Analytical Chemistry: Fundamentals, Advances, and Perspectives, Anal. Chem., 86 (2014) 262-285.

[16] Trujillo-Rodríguez M. J, Nan H, Varona M, Emaus M. N, Souza I.D, Anderson J.L, Advances of Ionic Liquids in Analytical Chemistry. Anal. Chem. 2019, 91: 505-531.

[17] Armstrong D.W, He L.F, Liu Y.S, Examination of ionic liquids and their interaction with molecules, when used as stationary phases in gas chromatography. Anal. Chem. 1999, 71: 3873-3876.

[18] Anderson J. L, Ding J, Welton T, Armstrong D. W, Characterizing Ionic Liquids on the Basis of Multiple Solvation Interactions. J. Am. Chem. Soc. 2002, 124: 14247-14254.

[19] Anderson J. L, Ding R, Ellern A, Armstrong D. W, Structure and properties of high stability geminal dicationic ionic liquids. J. Am. Chem. Soc. 2004, 127: 593-604.

[20] Sharma P. S, Payagala T, Wanigasekara E, Wijeratne A. B, Huang J, Armstrong D. W, Trigonal Tricationic Ionic Liquids: Molecular Engineering of Trications to Control Physicochemical Properties. Chem. Mater. 2008, 20: 4182-4184.

[21] Breitbach Z. S, Armstrong D. W, Characterization of phosphonium ionic liquids through a linear solvation energy relationship and their use as GLC stationary phases. Anal. Bioanal. Chem. 2008, 390: 1605-1617.

[22] Davis J. H, Task-Specific Ionic Liquids. Chem. Letters. 2004, 33: 1072-1077.

[23] Huang K, Han X, Zhang X, Armstrong D, PEG-linked geminal dicationic ionic liquids as selective, high-stability gas chromatographic stationary phases. Anal. Bioanal. Chem. 2007, 389: 2265-2275.

[24] López-Martin I, Burello E, Davey P.N, Seddon K.R, Rothenberg G, Anion and Cation Effects on Imidazolium Salt Melting Points: A Descriptor Modelling Study. ChemPhysChem, 2007, 8: 690-695.

[25] Murray S.M, O'Brien R.A, Mattson K.M, Ceccarelli C, Sykora R.E,West K.N, Davis J.H, The Fluid-Mosaic Model, Homeoviscous Adaptation, and Ionic Liquids: Dramatic Lowering of the Melting Point by Side-Chain Unsaturation, Angew Chem Int Edit. 2010, 49: 2755-2758.

[26] Mirjafari A, O'Brien R. A, West K.N, Davis J.H, Synthesis of New Lipid-Inspired Ionic Liquids by Thiol-ene Chemistry: Profound Solvent Effect on Reaction Pathway, Chemistry-a. European Journal. 2014, 20: 7576-7580.

[27] Dhanesar S.C, Coddens M.E, Poole C.F, Influence of phase loading on the performance of whisker-walled open tubular columns coated with organic molten salts. J. Chromatogr. A. 1985, 324: 415-421.

[28] Poole C. F, Furton K.G, Kersten B.R, Liquid Organic Salt Phases for Gas Chromatography. J. Chromatogr. Sci. 1986, 24: 400-409.

[29] Seeley J.V, Seeley S.K, Multidimensional gas chromatography: fundamental advances and new applications. Anal. Chem. 2013, 85: 557-578.

[30] Marriott P.J, Chin S.T, Maikhunthod B, H.-G. Schmarr H.-G, Bieri S, Multidimensional gas chromatography Trac-Trends. Anal. Chem. 2012, 34: 1-21.

[31] Dimandja J. M.D, Peer Reviewed: Comprehensive 2-D GC provides high-performance separations in terms of selectivity, sensitivity, speed, and structure. Anal. Chem. 2004, 76: 167 A-174 A.

[32] Zakaria M, Gonnord M. F, Guiochon G, Applications of two-dimensional thin-layer chromatography. J. Chromatogr. A 1983, 271: 127-192.

[33] Marriott P. J, Chin S. T, Maikhunthod B, Schmarr H. G, Bieri S, Multidimensional gas chromatography. Trends Anal. Chem. 2012, 34: 1-21.

[34] Simmons M. C, Snyder L. R, Two-stage Gas liquid chromatography. Anal. Chem. 1958, 30, 32-35.

[35] Seeley J. V, Bates C. T, McCurry J. D, Seeley S. K, Stationary phase selection

and comprehensive two-dimensional gas chromatographic analysis of Trace biodiesel in petroleum-based fuel. J. Chromatogr. A 2012, 1226: 103-109.

[36] Ragonese C, Sciarrone D, Tranchida P.Q, Dugo P, Mondello L, Use of ionic liquids as stationary phases in hyphenated gas chromatography techniques. J. Chromatogr. A 2012, 1255: 130-144.

[37] Krupčík J, Gorovenko R, Špánik I, Bočková I, Sandra P, Armstrong D. W, On the use of ionic liquid capillary columns for analysis of aromatic hydrocarbons in low-boiling petrochemical products by one-dimensional and comprehensive two-dimensional gas chromatography. J. Chromatogr. A. 2013, 1301: 225-236.

[38] Seeley J. V, Seeley S. K, Libby E. K, Breitbach Z. S, Armstrong D. W, Comprehensive two-dimensional gas chromatography using a high-temperature phosphonium ionic liquid column. Anal. Bioanal. Chem. 2008, 390: 323-332.

[39] Siegler W. C, Crank J. A, Armstrong D. W, Synovec R. E, Increasing selectivity in comprehensive three-dimensional gas chromatography via an ionic liquid stationary phase column in one dimension. J. Chromatogr. A 2010, 1217: 3144-3149.

[40] Nosheen A, Mitrevski B, Bano A, Marriott P. J, Fast comprehensive two-dimensional gas chromatography method for fatty acid methyl ester separation and quantification using dual ionic liquid columns. J. Chromatogr. A 2013, 1312: 118-123.

[41] Rodríguez-Sánchez S, Galindo-Iranzo P, Soria A. C, Sanz M. L, Quintanilla- López J. E, Lebrón-Aguilar R, Characterization by the solvation parameter model of the retention properties of commercial ionic liquid columns for gas chromatography J. Chromatogr. A 2014, 1326: 96-102.

[42] Yuan J.Y, Mecerreyes D, Antonietti M, Poly (ionic liquids): An update. Prog. Polym. Sci. 2013, 38: 1009-1036.

[43] Lin H, Zhang S, Sun J.-K, Antonietti M, Yuan J, Poly (ionic liquids) with engineered nanopores for energy and environmental applications. Polymer. 2020, 202: 122640.

[44] Zulfiqar S, Sarwar M.I, Mecerreyes D, Polymeric ionic liquids for CO2 capture and separation: Potential,progress and challenges. Polym. Chem. 2015, 36: 6435-6451.

[45] Morinaga T, Honma S, Ishizuka T, Kamijo T, Sato T, Tsujii Y, Synthesis of Monodisperse Silica Particles Grafted with Concentrated Ionic Liquid-Type Polymer Brushes by Surface-Initiated Atom Transfer Radical Polymerization for Use as a Solid State Polymer Electrolyte. Polymers 2016, 8: 146.

[46] Gupta N, Liang Y.N, Hu X, Thermally responsive ionic liquids and polymeric ionic liquids:emerging trends and possibilities. Curr. Opin. Chem. Eng. 2019, 25: 43-50.

[47] García-Verdugo E, Altava B, Burguete M.I, Lozano P, Luis S.V, Ionic liquids and continuous flow processes: A good marriage to design sustainable processes. Green Chem. 2015, 17: 2693-2713.

[48] Escorihuela J, Olvera-Mancilla J, Alexandrova L, Del Castillo L. F, Compan V, Recent Progress in the Development of Composite Membranes Based on Poly benzimidazole for High Temperature Proton Exchange Membrane (PEM) Fuel Cell Applications. Polymers. 2020, 12, 1861.

[49] Villa R, Alvarez E, Porcar R, Garcia-Verdugo E, Luis S.V, Lozano P, Ionic liquids as an enabling tool to integrate reaction and separation processes. Green Chem. 2019, 21: 6527-6544.

[50] Patinha D.J.S, Wang H, Yuan J, Rocha S.M, Silvestre A.J.D, Marrucho I.M, Thin Porous Poly(ionic liquid) Coatings for Enhanced Headspace Solid Phase Micro extraction. Polymers 2020, 12: 1909.

[51] Tian Y, Feng X.L, Zhang Y.P, Yu Q, Wang X.H, Tian M.K, Determination of VolatileWater Pollutants Using Cross-Linked Polymeric Ionic Liquid as Solid Phase Micro-Extraction Coatings. Polymers. 2020, 12: 292.

[52] Nan H, Anderson J.L, Ionic liquid stationary phases for multidimensional gas chromatography. Trends Anal. Chem. 2018, 105: 367-379.

[53] Trujillo-Rodríguez M.J, Nan H, Varona M, Emaus M.N, Souza I.D, Anderson J.L, Advances of ionic liquids in analytical chemistry. Anal Chem. 2019, 91: 505-531.

[54] Gónzalez-Álvarez J, Blanco-Gomis D, Arias-Abrodo P, Díaz-Llorente D, Busto E, Ríos-Lombardía N, Gotor-Fernández V, Gutierrez-Álvarez M.D, Polymeric imidazolium ionic liquids as valuable stationary phases in gas chromatography: Chemical synthesis and full characterization. Anal. Chim. Acta. 2012, 721: 173-181.

[55] Poole C.F, Poole S.W, Ionic liquid stationary phases for gas chromatography. J. Sep. Sci. 2011, 34: 880-900.

[56] Cagliero C, Mazzucotelli M, Rubiolo P, Marengo A, Galli S, Anderson J.L, Sgorbini B, Bicchi C, Can the selectivity of phosphonium based ionic liquids be exploited as stationary phase for routine gaschromatography? A case study: The use of trihexyl(tetradecyl) phosphonium chloride in the flavor, fragrance and natural product fields. J. Chromatogr. A 2020, 1619: 460969.

[57] Twu, P, Zhao Q, Pitner W.R, Acree W.E, Baker G.A, Anderson J.L, Evaluating the solvation properties of functionalized ionic liquids with varied cation/anion composition using the solvation parameter model. J. Chromatogr. A 2011, 1218: 5311-5338.

[58] Rodríguez-Sánchez S, Galindo-Iranzo P, Soria A.C, Sanz M.L, Quintanilla-López J.E, Lebrón-Aguilar R, Characterization by the solvation parameter model of the retention properties of commercial ionic liquid columns for gas chromatography. J. Chromatogr. A 2014, 1326: 96-102.

[59] Shashkov M.V, Sidelnikov V.N, Orthogonality and Quality of GC_GC Separations for Complex Samples with Ionic Liquid Stationary Phases in First Dimension. Chromatographia. 2019, 82: 615-624.

[60] Poole C.F, Lenca N, Gas chromatography on wall-coated open-tubular columns with ionic liquid stationary phases. J. Chromatogr. A. 2014, 1357: 87-109.

[61] Odugbesi G.A, Nana H, Soltani M, Davis J.H. Jr, Anderson J.L, Ultra-high thermal stability perarylated ionic liquids as gas chromatographic stationary phases for the selective separation of polyaromatic hydrocarbons and polychlorinated biphenyls. J. Chromatogr. A. 2019, 1604: 460-466.

[62] Roeleveld K, David F, Lynen F, Comparison between polymerized ionic liquids synthesized using chain-growth and step-growth mechanisms used as stationary phase in gas chromatography. J. Chromatogr. A. 2016, 1451: 135-144.

[63] Zhao Q, Anderson J.L, Highly selective GC stationary phases consisting of binary mixtures of polymeric ionic liquids. J. Sep. Sci. 2010, 33: 79-87.

[64] Ho W-Y, Hsieh Y.-N, Lin W.-C, Kao C.L, Huang P.C, Yeh C.F, Pane C.Y, Kuei C.H, High temperature imidazolium ionic polymer for gas chromatography. Anal. Methods. 2010, 2: 455-457.

[65] González-Álvarez J, Arias-Abrodo P, Puerto M, Viguri M.E, Pérez J, Gutiérrez-Álvarez M.D, Polymerized phosphonium-based ionic liquids as stationary phases in gas chromatography: Performance improvements by addition of graphene oxide. New J. Chem. 2015, 39: 8560-8568.

[66] Devanathan R, Recent developments in proton exchange membranes for fuel cells. Energy Environ.Sci.2008, 1:101-119.

[67] Quartarone E, Mustarelli P, Electrolytes for solid-state lithium rechargeable batteries: recent advances and perspectives. Chem.Soc.Rev.2011, 40: 2525-2540.

[68] Lissner E, De Souza W.F, Ferrera B, Dupont J, Oxidative Desulfurization of Fuels with Task-Specific Ionic Liquids. Chem.Sus.Chem. 2009, 2: 962-964.

[69] Fei Z.F, Zhao D.B, Geldbach T. J, Scopelliti R, Dyson P.J, Brønsted Acidic Ionic Liquids and Their Zwitterions: Synthesis, Characterization and pKa determination. Chem.Eur.J. 2004, 10: 4886-4893.

[70] Lin J.C.Y, Huang C.J, Lee Y.T, Lee K.M, Lin I.J.B, Carboxylic acid functionalized imidazolium salts: sequential formation of ionic, zwitterionic, acid-zwitterionic and lithium salt-zwitterionic liquid crystals. J. Mater. Chem. 2011, 21: 8110-8121.

[71] Nockemann P, Thijs B, Pittois S, Thoen J, Glorieux C, Van Hecke K, Van Meervelt L, Kirchner B, Binnemans K. Task-Specific Ionic Liquid for Solubilizing Metal Oxides. J. Phys. Chem. B 2006, 110: 20978-20992.

[72] Kuroda K, Satria H, Miyamura K, Tsuge Y, Ninomiya K, Takahashi K, Design of wall-destructive but membrane-compatible solvents. J. Am. Chem. Soc. 2017, 139: 16052-16055.

[73] Yoshizawa-Fujita M, Tamura T, Takeoka Y, Rikukawa M, Low-melting zwitterion: effect of oxyethylene units on thermal properties and conductivity. Chem. Commun., 2011, 47: 2345-2347.

[74] Straathof A. J. J, Transformation of biomass into commodity chemicals using enzymes or cells. Chem. Rev. 2014, 114: 1871-1908.

[75] Lange J.P, Price R, Ayoub P.M, Louis J, Petrus L, Clarke L, Gosselink H, Valeric biofuels: a platform of cellulosic transportation fuels. Angew. Chem. Int. Ed. 2010, 49: 4479-4483.

[76] Bond J.Q, Alonso D.M, Wang D, West R.M, Dumesic J.A, Integrated catalytic conversion of γ-valerolactone to liquid alkenes for transportation fuels. Science. 2010, 327: 1110-1114.

[77] Ragonese C, Tranchida P.Q, Sciarrone D, Mondello L, Conventional and fast gas Chromatography analysis of biodiesel blends using an ionic liquid stationary phase. J.Chromatogr. A 2009, 1216: 8992-8997.

[78] Baltazar Q.Q, Leininger S.K, Anderson J.L, Binary ionic liquid mixtures as gas chromatography stationary phases for improving the separation selectivity of alcohols and aromatic compounds. J. Chromatogr. A. 2008, 1182: 119-127.

[79] Joshi M. D, Anderson J. L, Recent advances of ionic liquids in separation science and mass spectrometry. RSC Adv. 2012, 2: 5470-5484.

[80] Anderson J. L, Armstron D. W, Wei G.T, Ionic liquids in Analytical Chemistry. Anal. Chem. 2006, 78: 2892-2902.

[81] Wang Y, Tian M, Bi W, Row K. H, Application of Ionic Liquids in High Performance Reversed-Phase Chromatography. I.J.M.S. 2009, 10: 2591-2610.

[82] Han X, Armstrong D. W, Ionic Liquids in Separations. Accounts Chem. Res. 2007, 40: 1079-1086.

[83] Tang S, Baker G. A, Zhao H, PEG-functionalized ionic liquids for cellulose dissolution and saccharification. Chem. Soc. Rev. 2012, 41: 4030-4066.

[84] Plechkova N. V, Seddon K. R, Applications of ionic liquids in the chemical industry. Chem. Soc. Rev. 2008, 37: 123-150.

[85] Cagliero C, Bicchi C, Ionic liquids as gas chromatographic stationary phases: how can theychange food and natural product analyses?, Analytical and Bioanalytical Chemistry 2020, 412: 17-25.

Chapter 6

Ancient and Contemporary Industries Based on Alkali and Alkali-Earth Salts and Hydroxides: The Historical and Technological Review

Rina Wasserman

Abstract

Although sodium, potassium, calcium, and magnesium were isolated *as the chemical elements* by Sir Humphry Davy for the first time at the beginning of the 19th century, alkali salts and hydroxides have been widely known and used since the very ancient time. The word "alcali" & "alkali" was borrowed in the 14th century by literary Roman-Germanic languages from Arabic *al-qalī*, *al-qâly* ou *al-qalawi* (القلوي), which means "calcinated ashes" of saltwort plants. These ashes are characterized nowadays as mildly basic. They have been widely used in therapy, cosmetics, and pharmacy in Mediaeval Europe and the Middle East. However, the consumption of these alkali containing ashes, as well as natron salts and calcined lime-based materials used for different customer purposes, like therapy, pharmacy, cosmetics, glass making, textile treating, dyes, brick making, binding materials, etc., was commonly known since the very ancient times. The current review of the archeological, historical, and technological data provides the readers with the scope of the different everyday life applications of alkali and alkali-earth salts and hydroxides from ancient times till nowadays. The review obviously reveals that many modern chemical manufacturing processes using alkali and alkali-earth salts and hydroxides have a very ancient history. In contrast, there has been a similarity of targets for implementing alkali and alkali-earth salts and hydroxides in everyday life, from the ancient past till the modern period. These processes are ceramic and glass making, binding materials in construction, textile treatment, metallurgy, etc. So, this review approves the common statement: "The Past is a clue for the Future."

Keywords: alkali, caustic, lime, pH, natural cement, Portland cement

1. Introduction

The alkali metals Na and K reside in the first column of the periodic table. Sir Humphry Davy, a prominent English scientist of the 19th century, electrolyzed sodium, Na, and potassium, K, and named them in 1807 [1]. At first, Davy called metallic potassium and sodium "the basis of potash" and "basis of soda," respectively. Consistently, he renamed these new metals potassium and sodium. Dmitry Mendeleev designated this discovery as "… one of the greatest discoveries in Chemistry … " [2]. However, after discovering these alkali elements that play an essential role in modern life, they have been little known by non-scientists for many years [3]. Although potassium and sodium as metals entered human life only ca. 200 years ago, humans have been familiar with their substances for thousands of years. Usage, treatment, and conscious transformations of alkali substances, which are almost six thousand years old, could be called the first advanced chemical technology in humankind's Big history.

The word "alcali" & "alkali" was borrowed in the 14th century by literary Roman-Germanic languages from Arabic al-qalī, al-qâly ou al-qalawi (القلوي), which means "calcinated ashes" of saltwort plants. These ashes are chemically characterized nowadays as mildly basic. They have been widely used in therapy, cosmetics, and pharmacy in Mediaeval Europe and the Middle East. However, the consumption of these alkali containing ashes, as well as natron salts and calcined lime-based materials used for different customer purposes, like therapy, pharmacy, cosmetics, glass making, textile treating, dyes, brick making, binding materials, etc., was commonly known since the very ancient times.

The current article intends to review those technological processes of alkali substances, modernly called 'chemical technology,' and track these processes' ancient and historical roots revealed by archeological findings and historical descriptions.

Undoubtedly, the ancient civilizations were not aware of the contemporary "chemical language" and did not carry out any scientific investigations or testing the chemical and technological procedures before their implementation. However, the archeological findings have revealed in the last 100 years a vast amount of built-in chemical knowledge possed by the prominent ancient civilizations in their everyday life. Alkali salts played an essential role in human health and body care during ancient times. The ancient texts' interpretations have revealed alkalis substances' conscious usage as detergent and hygienic remedy throughout human history. Furthermore, the ancient texts have distinguished between alkaline salts' mineral and botanical origin, although emphasizing similar usage. Let us get down to some examples of alkali-based substances' knowledge and use in ancient and historical times.

2. Use of alkaline salts in ancient and historic cosmetics, food, cleaning, and medicine

2.1 Alkaline salts as the most initial raw materials of the ancient Mesopotamian pharmacology

The first documented use of ordinary table salt and soda could be related to Sumer and Akkadian Empires in Mesopotamia (3500–2000 BCE). **Figure 1** presents the map of the ancient Near East in the fourth millennium DC. At the beginning of the 20th century, the University Museum's archeological expedition, Philadelphia, the USA, to Nippur (a lower part of modern Iraq) excavated the cuneiform tablet aged ca. 2100 BCE [5]. In the tablet (**Figure 2**) decrypted at the half of the 20th century, the Sumerian script described the pharmacological processes involving alkaline substances of mineral and botanical origin. The mineral salts of alkali metals mentioned in the tablet are sodium chloride and potassium nitrate. Sumerians obtained alkalis also from soda ash, which they called Td-Gaz. This soda ash had a botanical origin by burning halophytic (high salinity) and alkaline plants, like glassworts (most likely the *Salicornia fruticosa* L.) rich in sodium carbonate, Na_2CO_3

Figure 1.
Map of ancient Mesopotamia [4].

[5, 7]. Sumerians had an abundance of designations for these alkaline plants and alkali ashes and salts according to their origin and manufacturing processes [8–10]:

- *di-ni-ig*: potash; salt (dè, ‘ashes’, + nağa, ‘potash’)
- *nağa*: potash, K_2CO_3, soap;
- *nağa-Gaz*: crushed soda (plant?) (‘soda, alkali’ + ‘to crush, to grind, to grate, powder’);
- *nağa-si-è*: sprouted alkaline plant (‘soapwort’ + ‘antennae’ + ‘to go forth’);
- *ne-mur*: glowing coals, fire, alkali, potash; ashes; charcoal (“tree” + compound verb verbal element)

Glassworts are hardy to high alkaline environments and store absorbed alkaline salts in their tissues during growth [11–13]. According to the modern analytical tests, sodium and potassium carbonate content in ashes obtained from the Near East halophytic plants could be 38.5% - 93% [11]. According to the Sumerian script, the further treatment of soda ash included its multistage purification and pulverization processes. Sumerians widely implemented this “halophytic ash” soda in pharmacology, making simple detergents and soaps for body cleansing and religious purifying:

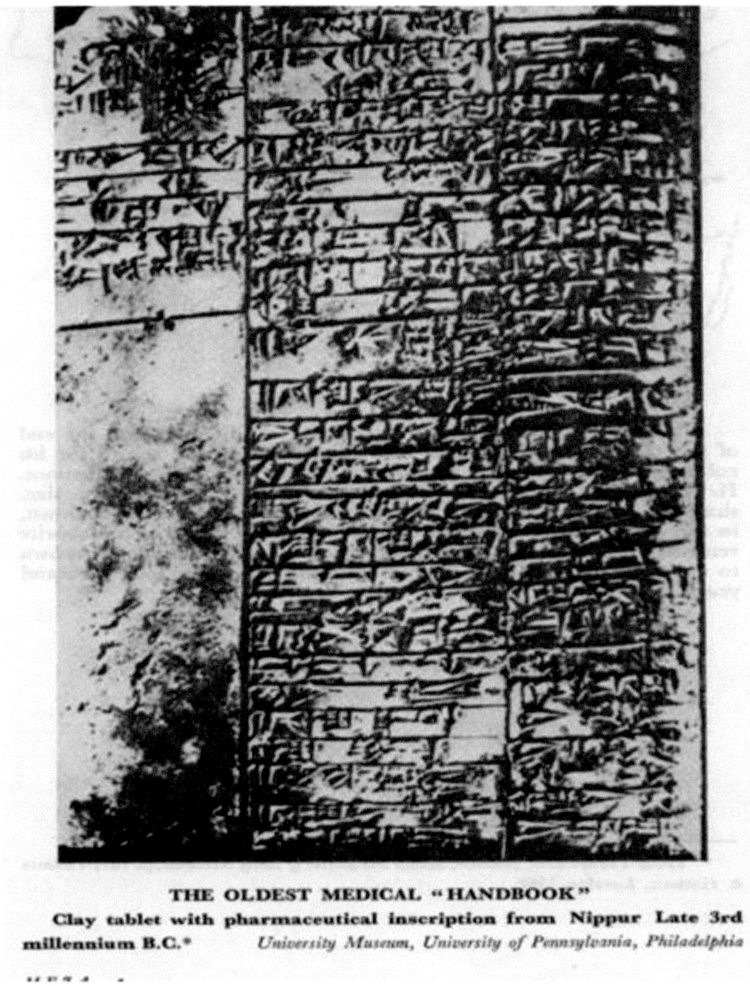

Figure 2.
Picture of cuneiform (clay tablet) with the pharmacological inscription, Nippur, c. 2100 B.C. University of Pennsylvania, Philadelphia USA. [6].

"With water I bathed myself. With soda I cleansed myself. With soda from a shiny basin I purified myself." [14].

2.1.1 The rise of soapmaking

Another Sumerians' pharmacological use of sodium salts (soda ash and regular salt, NaCl) was making a medicated ointment soap as a rubbing remedy for ailments. The preparation process was based on thorough mixing of sodium salts, i.e., sodium ash and regular salt, with various natural organic ingredients. [7]. Generally, the earliest Sumerians' soaps were made for medical purposes and wool washing but not for general cleansing purposes. To extract alkali from the plants, Sumerians put into use the following technological stages [14]:

- slow combustion or incineration of the dried halophytic and alkaline plants;
- leaching or washing of the plant ash;

- separating the mixture by water evaporation and drying until the salt cake crystallizes on the vessel's sides;

- calcination of the crude product to ignite organic substances.

This method was based on a long-time, slow and thorough process to assure the high yield of alkalis' extraction and, therefore, the more expensive product was obtained. The Sumerian elites used this product for ritual purifying. Common Sumerians got a more simple leaching process in everyday life. They stirred the plant ash in water and filtered the suspension before using it to remove the insoluble impurities. The resulting basic lixivium (alkali leachate) was widely used for everyday cleansing and washing purposes.

2.1.2 Advanced technologies for table salt manufacturing

The Old and New Babylonian languages, which are Sumerian and Akkadian, respectively, used the same logogram for table salt , [9, 10, 15], but had the different pronunciation of this term: *mu-n*(u) in Sumerian (Old Babylonian) vs. *ṭab-tu(m)* in Akkadian (New Babylonian) [8, 16].

Table salt, which is sodium chloride, NaCl, is described in the decrypted Sumerian (Old Babylonian) texts as an essential ingredient of the human's diet, food preservative, and the pharmacological salting-out ingredient to separate a medicated ointment soap from the glycerin, excess of water, and impurities [7, 14, 17, 18].

A great deal of salt treatment in various parts of the Ancient World, since the prehistoric times, till the 19th century C.E., was the uniformity of salt production techniques [19]:

a. Rock salt was mined or gathered.

b. Seawater and brine were universally used to remove salt by natural evaporation or artificial solution boiling, respectively.

c. Halophytic plant ashes were washed and refined.

d. Salt molds from different places worldwide were based on very similar ceramic vessels.

The utilization of salines for salt production was probably the most crucial method of salt superiority in antiquity. The leather sacks were used in Ancient Mesopotamia to transport large quantities of salt from salines to villages and towns. Also, salt molds made of porous ceramic material or reeds were used to transport the precipitated "salt cakes" to consumers grinding them for daily use. Wooden bowls served as salt containers in the home [17]. In the Sumerian and Akkadian Empires, salt served as a reward for work as a state servant. According to the decrypted Akkadian texts, unskilled workers and high-quality artisans employed at the reconstruction of the Ékur Temple (**Figure 3**) in Nippur under the Kings Naram-Sin (reigned 2261–2224 BCE) and Shar-Kali-Sharri (ruled 2217–2193 BCE) were given 0.421 and 0.842 liters of salt per capita per month, respectively [20].

Those amounts' mass equivalent comprises an average of 10.4 and 20.8 gr salt per person per day in a 30-day month, respectively [17]. According to the U.S. Food and Drug Administration, the recommended daily sodium intake is less than

Figure 3.
Photograph of Ekur, the ziggurat of Enlil at Nippur [4].

2,300 mg per day [21]. In the context of salt intake, one gram of sodium equals approximately 2.5 gr of salt [22]. Thus, by present-day measurement, the ancient Accadian workers obtained a surplus of salt as salary. Perhaps, this was done because of the extensive use of salt as an animal food' preservative in a hot climate. Also, salt played a high role in many medical remedies to help afflictions of the soul, psyche, male virility, and magic rituals in the Ancient Sumer and Accadian empires [17]. Considering the Akkadian Empire's high population during the third millennium BCE, massive salt consumption could be imagined. Therefore, large-scale logistics of salt gathering, a well-established delivery system, massive salt loading and unloading operations, commodity distribution, and numerous material equipment could be considered for that ancient society [17].

2.1.3 Saltpeter green manufacturing

Another alkali substance described by Sumerians pharmacological text [5] was niter or saltpeter, K_2NO_3. Sumerians called saltpeter *mú-nu* (*mú*, 'to make grow,' + *nu*, 'fire'), whereas the Akkadian word for saltpeter was *marru* (bitter) [17]. Nevertheless, also, in this case, they used for this alkaline salt the same logogram  [10]. The Old Babylonian (Sumerian) word used for saltpeter in the 4th millennium DC demonstrates the knowledge of potassium nitrate ignitability. It should be emphasized that the last millennium civilization has begun using the saltpeter as the oxygenating ingredient in gunpowder only since 9th century AD [23, 24].

An exciting fact is that Sumerians and Assyrians obtained this salt by a crystalline formation from the surface drains containing nitrogenous urine waste products. This intermedial crystallized substance contained a mixture of alkali salts (sodium chloride, potassium nitrate, and others). According to this ancient script, Sumerians harnessed fractional crystallization processes to purify obtained niter. The text, although, emphasized the yield of the processes. Our days, we call such processes of 'green' innovative and emerging technologies [25–29].

2.2 Descendants of the knowledge and experience of sodium alkaline salts

The knowledge and experience of sodium alkaline salts' use were "inherited" by Babylon (2000–144 BCE) and the Hittite (1900–700 BCE) empires (both in Mesopotamia). Alkaline substances based on wood or plant ash were in high demand there. A ca.1300 BCE text described the Hittites using sodium salt, possibly sodium soda, and another alkali compound from plants, for washing the hands in a religious practice [14]. In ancient Babylonia, many halophytic plants were used for their alkaline substances. Babylonians used the same operations as Sumerians to extract alkalis from the plants for cleansing, hygienic, ritual, cosmetical, and medical purposes.

The usage of table salt, NaCl, as a food preservative agent was well known in ancient Mesopotamia in the first millennium BCE [18]. Table salt was an essential product in the food allowance paid by Ancient Mesopotamian rulers to workers and was considered a necessity for regular human maintenance.

2.3 Alkaline salts in the ancient Egyptian pharmacology (3000 BCE: 395 CE)

Another civilization distinguished by a conscious understanding of alkalis salts' importance and their broad medical and pharmacological usage was Ancient Egypt (3000 BCE – 395 CE). Alkalis salts used by the Egyptians since very ancient times were mineral soda (natron), $Na_2CO_3{\cdot}NaHCO_3{\cdot}xH_2O$, regular salt, NaCl, and sodium sulfate, Na_2SO_4 [30–32]. Natural Soda occurs in Egypt principally in the Wadi Natrun in the Libyan desert (**Figure 4**), and to less extent, at El Barnugi, in Lower Egypt, and at Mahamid, in Upper Egypt.

The Wadi Natrun deposits have been probably the oldest known source of natural soda globally, and they served to supply that commodity for thousands of

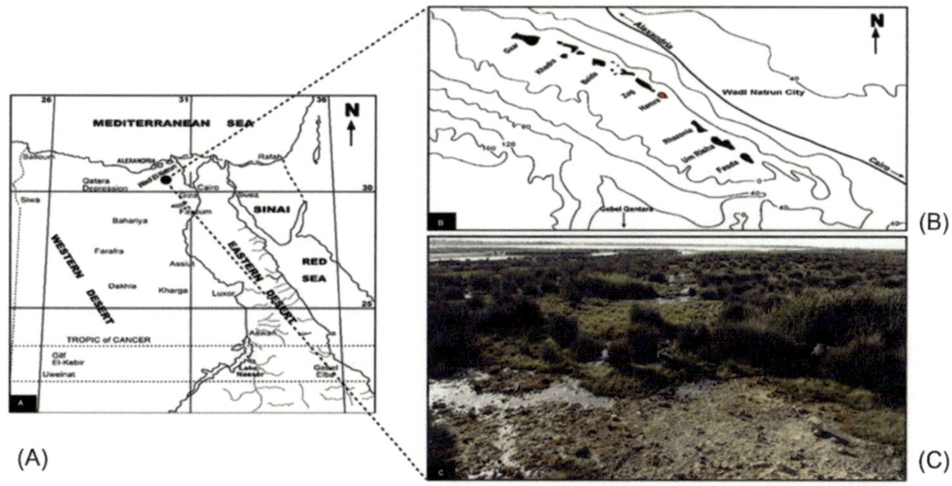

Figure 4.
(A) Location map of Wadi El Natrun (Western Nile Delta), (B) inland saline lakes including Lake Hamra at Wadi El Natrun, (C) the swamp beside Lake Hamra about 5 m above the lake level. (picture copyright © [33]).

years. In ancient times there were two soda lakes, which became united when water was most abundant [30]. The soda was found there in three forms:

a. as a solution in the lakes' water;

b. in a solid form at the bottom of the lakes;

c. as a crust on the ground.

The archeological excavations of the 19th and 20th centuries revealed (among others) seventeen ancient Egyptian papyri dealing with medical, pharmacological, and body purification issues and covering the period of more than two millennia, dating from the Middle Kingdom (2040–1640 BCE) to the end of the Greco-Roman Period (332 BCE-395 CE) until the Roman Empire's break up [34]. These papyri proved the importance of sodium salts in Ancient Egypt. However, sodium salts' usage in medicine, mummification, and ancient Egypt's hygiene began much earlier, five millennia ago, during the Early Dynastic Period (2920–2575 BCE) [31, 32].

The papyri mentioned plenty of times usage of sodium-based minerals as medication commodity:

a. salt from seven geographic origins. The salt, which is sodium chloride, was mentioned 175 times in total. The Egyptians could have potentially exploited the local salt from the Wadi Natrun and the Oasis of Siwa, harvested from salt pans in the Mediterranean and imported into Egypt from the Sinai Peninsula and some other sea salt works abroad [31].

b. natron from five geographic origins. The use of natron was described in papyri 92 times in total. The primary sources of natron within Ancient Egypt were the deposits at Wadi Natrun in the Lybian desert and El Kab in Upper Egypt south to Luxor. Ancient Egypt also imported natron from Sudan, Nubia, and Syria [31].

Furthermore, the salt and natron's mentions are more than one-third of the papyri's total mineral references. No other mineral was mentioned so frequently in the medical. These figures confirm the importance of sodium salts for health treatments in Ancient Egypt.

According to the papyri, the most common causes of salt usage in Ancient Egypt were treating wounds and mummifications (sodium chloride is well known as a putrefaction inhibitor), cure against diarrhea and dehydration, and cosmetics [31]. The interesting fact is that the use of salt is still quite common among traditional folk remedies.

In Ancient Egyptian medicine, natron's use was very similar to salt, mostly externally for wound treatments, skin curing, purification, mummification and embalming, and washing. M. Sapsford carried out the analytical tests to reveal and estimate salt and natron's role in the papyri's skin-curing prescriptions [31]. This study showed that salt served a moisture's retainer's role in anti-wrinkle skin creams, and natron possessed the highly desiccant ability. For the purification, salt and natron were used after an illness as a ritual cleansing means to be re-accepted as a fully functioning person in Ancient Egyptian society. Natron was also used as a means of unique purifying oneself after a period of "uncleanliness." There is some papyri' evidence of salt and natron' use for laundry' state-provided service. The exceptional importance of salt and natron for Ancient Egyptian society could be demonstrated by the fact that the village's households received a part of their monthly ration package by salt and natron [35].

2.4 Standard practices with alkaline salts adopted by ancient Hebrews, Greeks, and Romans (ca. 1150 BCE: 500 CE)

The Old Testament texts describe the wide use of salt by ancient Hebrews for a variety of purposes [36]:

a. sacrifices;

b. with food;

c. medicinal properties as an anti-infective agent;

d. as a symbol of the permanence of a covenant due to the observation that salt does not undergo decay;

e. as a symbol of perpetual desolation.

Ancient Hebrews distinguished between alkali sodas of different origins. They called natron, the mineral sodium carbonate, *neter*. In comparison, the word *borit* was used for sodium carbonate of vegetable origin (halophytic plant ash). The Ancient Hebrews commonly used *neter* and *borit* as laundry and body-cleaning agents [37].

The ancient Greeks and Romans eventually adopted these ancient practices of alkali salts' usage as laundering and body-cleaning agents and distributed them on the European continent [37]. However, since 600 BCE, the above-mentioned technological practices were changed to obtain solid soap, whereas wood ash became the main alkali-containing constituent, and animal fats were the binders in solid soaps. Pliny the Elder attributed the invention of alkali containing solid soap to one of the northern Celtic tribes [38]. Galen (129–199/216 CE), Greek physician, writer, and philosopher, wrote: "Soap is made by cooking beef, she-goat, or wether fat, mixed in with lye and quicklime." [39].

2.5 The Middle East as the disseminator of the ancient pharmacological crafts and knowledge of alkaline salts to Western and Southern Europe (since 7th CE)

Around 700–800 CE, the craft industry of soapmaking containing alkali plant ashes, animal fats, and the different plant oils became abundant in the Western and Southern Mediterranean, especially in Italy and Spain. This fact was mentioned in detail by Abū Mūsā Jābir ibn Ḥayyān, the Arab savant who lived in the 8th-century C.E. [40].

It should be emphasized that the English word "alkali" was "borrowed" from the Arabic language in the Middle Ages. The Arabic word القلوي al-qaly is the most common word for ash obtained from the alkaline saltworts plants [41]. Since ancient times, the caustic ashes' specific properties were well-known among the Middle East population. The Middle Eastern craftsmen inherited this time-honored traditional knowledge of the Ancient civilizations, whereas the Crusaders and the Arabic-speaking merchants helped disseminate this craft knowledge in Mediaeval Europe.

2.6 Declining the use of the natural alkali-bearing minerals since the industrial revolution in Europe

Until the Industrial Revolution in the 18th and 19th centuries, the caustic plant ashes had a primary use in semy-boiled body cleansing soaps possessing mild alkali pH. The natural mineral natron was the raw material for caustic washing and laundry soaps. At the end of the 18th century, Nicolas Leblanc, a French chemist, and surgeon invented an industrial process of converting the ordinary table salt, NaCl, into sodium carbonate, Na_2CO_3, to address the growing demand of the traditional industries in soda [42, 43]. Since that time, the soapmaking craft based on natural alkali-bearing minerals was gradually ousted by growing industrial technologies with chemically obtained detergents as raw materials. Industrial cleaning products of most of the 20th century have proven themselves very effective detergents. However, these artificial chemicals were found as highly allergenic and causing other unintended deleterious effects. Thus, the current trend to turn back to the traditional soapmaking crafts using the alkali-containing plant ashes has become very popular in the last years [44–46]. Thus, one more time in human history, the ancient knowledge of body cleaning and washing agents containing the plant ashes has been resurfaced and rejuvenated, being back now in human life's place.

3. Soda and potash-based glassmaking

As was previously mentioned, Leblanc's invention of artificial soda production at the end of the 18th century addressed a growing need of the European population and industry for caustic raw materials used in a) textile manufacturing as a bleaching agent; b) glassmaking as a soda-lime flux; and c) soapmaking for saponification of fats and oils [42].

3.1 2500 Year-long Ascension of crockery glassmaking based on soda and potash

Glassmaking has used natron, which is natural soda, since very ancient times. The first regularly produced glass was made in Egypt and the Near East in the sixteenth century BC [47]. The numerous archeological excavations revealed intensely colored glass, simulating precious stones such as turquoise, carnelian, lapis lazuli, amethyst, obsidian, and others, produced during the Late Bronze Age (1600–1200 BCE) [48]. Manufacturing any glass needs fluxes acting as atoms' network modifiers [49]. The network modifiers in very ancient glasses were the alkali metals and the alkali earths. The alkali metals, particularly sodium and potassium, disrupt the atom's network structure in glass, lower the melting point, and compromise the general stability of the glass (*ibid*). Alkali earths, especially calcium, usually counteract this effect to a certain extent and stabilize the glass. Ancient and historical glasses are alkali-lime-silicate glasses because alkali carbonates, such as plant ashes and natron, were the critical raw materials consciously used by glassmakers [31, 43, 48, 50–64]. It is now widely accepted that during the Late Bronze Age, Soda and potash-rich plant ash enhanced by increased lime content was the primary flux additive used to make glass in the ancient Near East [48, 50, 64, 65].

The use of natron and trona, the natural sodium carbonates, in the glassmaking of the ancient world began to be evident at *circa* 1000 BC [48] and continued almost two millennia [31, 47, 48, 53, 56, 61, 62]. The primary source of natron, for ancient Near Eastern glass manufacturing since 1600–1200 BC, was Wadi-el-Natrun, in

Egypt [31, 48, 53, 61, 62, 64, 66]. However, Pliny the Elder mentioned in his *Natural History* the natron deposits from the al-Barnuj region in the Egyptian Nile Delta, the Lake Van in the eastern region of Turkey, and the al-Jabbul lakes in Syria resources used by the ancient Greek and Roman glassmakers [67]. However, the current archeological research on the glass production in the ancient Near East in 1000 BC – 1000 AC has not provided an unambiguous opinion regarding the possibility of the ancient large-scale exploration of natron from the deposits other than Wadi-el-Natrunor al-Burnuj [54, 61]. Since the Roman era and till the 9th century BC, almost only Egyptian natron deposits supplied the flux raw material for global glassmaking. From the 7th century AC towards the end of the first millennium, the Old World's glassmaking crafts faced a shortage of mineral natron from Egypt and the Levant [51]. This natron shortage led to "re-inventing" the millennia-old alkali flux, i.e., glassmaking in Mediaeval Europe widely adopted the plant potash-ash fluxes [55, 57, 60, 63]. In the 9th and 10th centuries, the art of poly- and monochromatic luster-stained glass became very popular in the Near-Eastern Islamic world [68], see **Figure 5**.

The soda-containing flux used in Egyptian luster-stained glasses of the Islamic period (9th – 10th centuries) was natron, whereas luster-stained glass vessels from the Syria–Palestine region and Mesopotamia were crafted with sodium and potassium rich plant ashes [70].

3.2 Alkali-containing plant ashes catalyzed the invention of colored stained glass for architectural purposes

Colored stained glass has played a significant role in European architecture since the 12th century AC [60, 71, 72], see **Figure 6**.

From the 12th century up to 1440 AC, the European window glassmaking technique was a broad glass method for producing small rectangular glass sheets

Figure 5.
The ceramic dish with blue, green, and manganese-purple glaze, from Raqqa, northern Syria, 12th century AD. Presented in British museum (museum inventory number 1923.2–17.1). Picture-copyright ©discover Islamic art (MWNF) *[69].*

Figure 6.
A medieval window at Troyes cathedral, France (14th century). Wikipedia [73].

[43, 74]. Finally, from the 15th century until the mid-19th century, the primary window glassmaking technique in Europe was the crown glass method of producing sheet glass [74], see **Figure 7**.

Both these techniques use alkali-containing plant ashes as a flux. However, the alkali-containing plant ashes used in glassmaking differed in the different European regions. In the Eastern and Southern Mediterranean regions, the soda-rich halophytic glassworts' ash, pure or blended with natron, was imported from the Syrian-Palestine region and Egypt and widely used in glassmaking since Mediaeval times, thanks to the commercial and technological interconnections with the Islamic East [76, 77].

Since the 13th century, the glassmakers enhanced the raw plant ashes by admixture with higher lime and magnesium oxide content to obtain the glass with good chemical stability and low thermal expansion [78].

In Central Europe in the 12th – 18th centuries, the glassmaking crafts widely used potash and soda-rich wood ashes as fluxes, whereas higher contents of lead oxide and lime in ash were found effective for enhancing the glass durability [74, 79, 80].

3.3 Alkaline salts of different origins have promoted the continuous development of glass technologies for construction purposes

Towards the 15th century, window glassmaking in Central Europe began using sea salt, NaCl, as an additive to ash flux to control the window glass composition affecting glass mechanical stability [79]. As an additive to wood ash flux, sea salt, soda, and niter were widely used to produce cylinder (hull) glass, a more advanced form of broad-glass manufacture, in Central and Nothern Europe until the 17th century (see **Figures 8** and **9**). The alternative important potash-containing material used as a flux in window glass manufacturing in England till the 19th century was kelp ash [82]. Kelp ash is a substance produced by the burning of seaweed [83]. The use of kelp ash in glass manufacturing was generally declined since the first half of the 19th century because of industrial sodium carbonate manufacturing (see **Figure 8**).

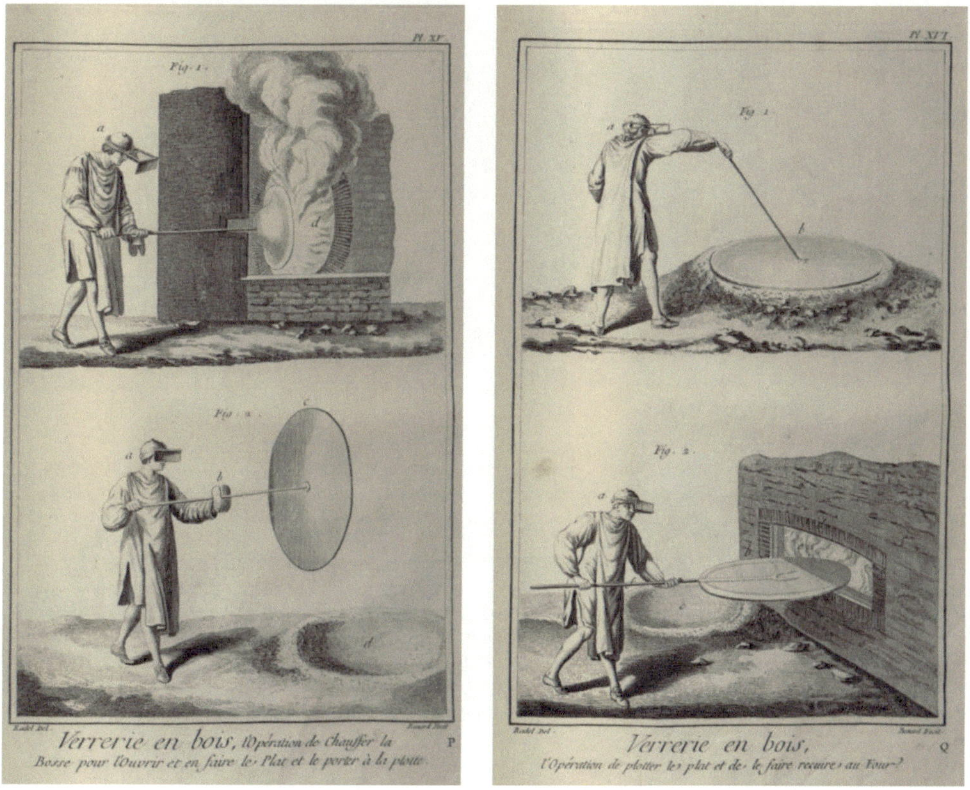

Figure 7.
*Robert Bénard (French artist, 1734–1777). Crown or window glass making engraving plates. Plates XV & XVI from [75]. The description (*in French*) of particular craft operations is also available online at [75].*

Figure 8.
Manufacturing cylinder (hull) glass. Engraving of a German glassworks, 1865. © Bildarchiv Preussischer Kulturbesitz *[81].*

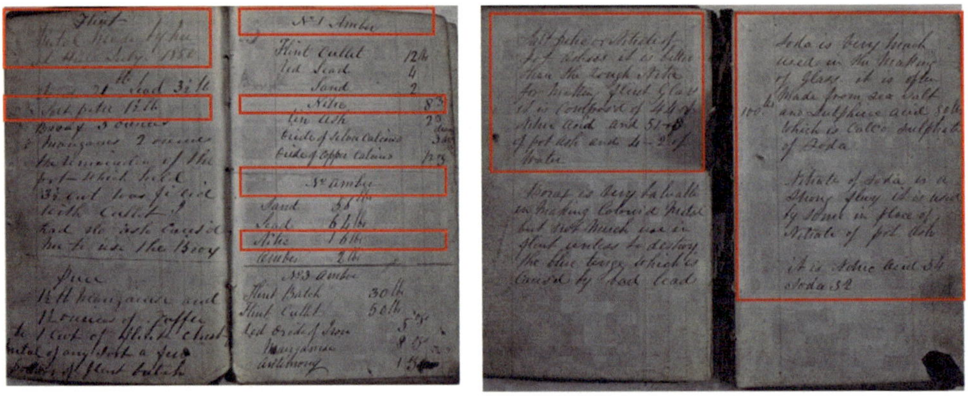

a) b)

Figure 9.
1850s Original Manuscript Book of glassmaking Recipes, procedures, and Formulaes. Pictures copyright © 2021 M. Benjamin Katz, fine books/rare manuscripts (https://www.mbenjaminkatzfinebooksraremanuscripts.com/product/4340/1850s-ORIGINAL-MANUSCRIPT-BOOK-OF-GLASSMAKING-RECIPES-PROCEDURES-AND-FORMULAES-UNIDENTIFIED). a) Text in the red rectangles on the left: *"Flint. Metal made by me at Hull July 1850 … Saltpeter 1½ lbs. … .".* Text in the red rectangles on the right: *"No. 1. Amber … nitre 8 oz. … . No. Amber … nitre 16 lbs". b)* Text in the red rectangles on the left: *"Saltpeter or nitrate of pot ashes it is better than the rough nitre for melting flint glass it is composed of 44% of nitric acid and 51% of pot ash and 4% of water".* Text in the red rectangles on the right: *Soda is very much used in the melting of glass it is often made from sea salt 100 lbs. and sulfuric acid 80 lbs. which is called sulphate of soda. Nitrate of soda is a strong flux it is used by some in place of nitrate of pot ash. It is nitric acid 54 soda 32."*

Using pure industrial carbonate of soda with sand and lime enabled the invention and manufacturing of a) large sheets of polished plate glass since the second half of the 19th century; b) drawn flat sheet glass from early in the 20th century; c) float glass since the late 1950s [74].

4. Conclusions

Alkali-containing salts have been the essential commodity in human life since very ancient times. Ancient civilizations studied to explore the natural resources of this commodity and developed sophisticated technological crafts and industries based on the specific properties of alkali-containing salts. For millennia this development was not based on scientific research and development but a trial-and-error approach. Nevertheless, the practical results of this empirical approach to the invention of the products based on alkali-containing salts were awe-inspiring.

Since the 19th century, synthetic alkali-containing carbonates have accelerated the industrial revolution in soap and washing detergents' production and window glass manufacturing. The invention of industrial alkali carbonate as a leading chemical commodity is fascinating because of the stages it went through; first, exploitation of natural resources for more than three and half millennia, followed by chemical industrial manufacturing for *ca.* one century, and a return to using the natural natron and plant ashes since the second half of the 20th century [84]. Turning back to the natural soda carbonate and potash resources since the late 90s of the 20th century is an obvious result of the crucial ecological approach to soap and glassmaking.

The Big History's holistic approach allows concluding that alkali-containing salts always have been essential for human well-being and highly appreciated raw materials. Thus, the millennia-old knowledge and use of this commodity have always been an authentic technological heritage.

Author details

Rina Wasserman
Department of Conservation of Sites and Monuments, Western Galilee College, Acre, Israel

*Address all correspondence to: rinaw@edu.wgalil.ac.il

References

[1] J. L. Dye, "The alkali metals: 200 years of surprises," Philos. Trans. R. Soc. A Math. Phys. Eng. Sci., vol. 373, no. 2037, p. 20140174, Mar. 2015.

[2] J. M. Thomas, P. P. Edwards, and V. L. Kuznetsov, "Sir Humphry Davy: Boundless chemist, physicist, poet and man of action," ChemPhysChem, vol. 9, no. 1, pp. 59–66, Jan. 2008.

[3] M. E. Weeks, "The discovery of the elements. IX. Three alkali metals: Potassium, sodium, and lithium," J. Chem. Educ., vol. 9, no. 6, pp. 1035–1045, Jun. 1932.

[4] The Oriental Institute and The University of Chicago, "Nippur - Sacred City Of Enlil Supreme God of Sumer and Akkad." [Online]. Available: https://oi.uchicago.edu/research/projects/nippur-sacred-city-enlil-2. [Accessed: 27-Jun-2021].

[5] M. Levey, "Ancient chemical technology in a Sumerian pharmacological tablet," J. Chem. Educ., vol. 32, no. 1, pp. 11–13, Jan. 1955.

[6] M. S. Takrouri, "Surgical, medical and anesthesia in the Middle East: Notes on Ancient and medieval practice with reference to Islamic-Arabic medicine," Internet J. Heal., vol. 5, no. 1, 2006.

[7] K. L. Konkol and S. C. Rasmussen, "An ancient cleanser: Soap production and use in antiquity," in ACS Symposium Series, vol. 1211, 2015, pp. 245–266.

[8] J. A. Halloran, Sumerian Lexicon. A Dictionary Guide to the Ancient Sumerian Language. Logogram Publishing, 2006.

[9] E. Peter, Sumerian Cuneiform English Dictionary. CreateSpace Independent Publishing Platform, 2014.

[10] Museum of Anthropology and Archaeology: Babylonian Section., "Pennsylvania Sumerian Dictionary," University of Pennsylvania, 2006. [Online]. Available: http://psd.museum.upenn.edu/. [Accessed: 27-Jun-2021].

[11] M. S. Tite, A. Shortland, Y. Maniatis, D. Kavoussanaki, and S. A. Harris, "The composition of the soda-rich and mixed alkali plant ashes used in the production of glass," J. Archaeol. Sci., vol. 33, no. 9, pp. 1284–1292, Sep. 2006.

[12] S. T. Ahmad, N. A. K. K. Sima, and H. H. Mirzaei, "Effects of sodium chloride on physiological aspects of salicornia persica growth," J. Plant Nutr., vol. 36, no. 3, pp. 401–414, Mar. 2013.

[13] D. Loconsole, G. Cristiano, and B. De Lucia, "Glassworts: From wild salt marsh species to sustainable edible crops," Agric., vol. 9, no. 1, p. 14, Jan. 2019.

[14] M. Levey, "The early history of detergent substances: A chapter in Babylonian chemistry," J. Chem. Educ., vol. 31, no. 10, pp. 521–524, Oct. 1954.

[15] Glosbe - The biggest online dictionary, "English-Akkadian dictionary." [Online]. Available: https://en.glosbe.com/en/akk/salt. [Accessed: 27-Jun-2021].

[16] J. Black, A. George, and N. Postgaye, Eds., A Concise Dictionary of Akkadian, 2nd (corre. Wiesbaden: Harrassowitz Verlag, 2000.

[17] D. Potts, "On Salt And Salt Gathering In Ancient Mesopotamia," J. Econ. Soc. Hist. Orient, vol. 27, no. 3, pp. 225–271, 1984.

[18] M. Levey, "Gypsum, Salt and Soda in Ancient Mesopotamian Chemical Technology," Isis, vol. 49, no. 3, pp. 336–342, Sep. 1958.

[19] J. A. E. Nenquin, "Salt. A Study in Economic Prehistory. Dissertationes Archaeologicae Gandenses, 6," De Tempel, vol. 6, 1961.

[20] J. Lines and D. J. Wiseman, The Alalakh Tablets, vol. 59, no. 4. London: British Institute of Archaeology at Ankara, 1955.

[21] FDA - U.S. Food and Drug Administration, "Sodium in Your Diet," 15/09/2020, 2012. [Online]. Available: https://www.fda.gov/food/nutrition-education-resources-materials/sodium-your-diet#:~:text=Americans eat on average about,about 1 teaspoon of salt! [Accessed: 05-Feb-2021].

[22] Joint Research Centre - European Commission, "Health promotion and disease prevention Knowledge Gateway - Whole Grain," Postgraduate Medicine, 2017. [Online]. Available: https://ec.europa.eu/jrc/en/health-knowledge-gateway/promotion-prevention/nutrition/whole-grain. [Accessed: 05-Feb-2021].

[23] J. R. Partington, A History of Greek Fire and Gunpowder. Baltimore, Mariland, USA: JHU Press, 1998.

[24] D. Cressy, Saltpeter: The Mother of Gunpowder. Oxford, UK: Oxford University Press, 2013.

[25] M. Maurer, P. Schwegler, and T. A. Larsen, "Nutrients in urine: Energetic aspects of removal and recovery," Water Sci. Technol., vol. 48, no. 1, pp. 37–46, Jul. 2003.

[26] M. Maurer, W. Pronk, and T. A. Larsen, "Treatment processes for source-separated urine," Water Res., vol. 40, no. 17, pp. 3151–3166, Oct. 2006.

[27] Z. Ganrot, G. Dave, and E. Nilsson, "Recovery of N and P from human urine by freezing, struvite precipitation and adsorption to zeolite and active carbon," Bioresour. Technol., vol. 98, no. 16, pp. 3112–3121, Nov. 2007.

[28] K. M. Udert and M. Wächter, "Complete nutrient recovery from source-separated urine by nitrification and distillation," Water Res., vol. 46, no. 2, pp. 453–464, Feb. 2012.

[29] C. M. Mehta, W. O. Khunjar, V. Nguyen, S. Tait, and D. J. Batstone, "Technologies to recover nutrients from waste streams: A critical review," Crit. Rev. Environ. Sci. Technol., vol. 45, no. 4, pp. 385–427, Feb. 2015.

[30] T. M. L, "Egyptian soda," Nature, vol. 90, no. 2254, pp. 527–528, Jan. 1913.

[31] Melanie Sapsford, "The use of sodium salt deposits in medical and medically associated industries in Ancient Egypt.," Cranfield University, 2009.

[32] A.-M. Gomaa and A. Elamin, "A review on the materials used during mummification processes in Ancient Egypt," Mediterr. Archaeol. Archaeom., vol. 11, no. 2, pp. 129–150, 2011.

[33] A. S. Zaky et al., "Mid-to Late Holocene paleoclimatic changes and paleoenvironmental shifts inferred from pollen and diatom assemblages at Lake Hamra, Wadi El Natrun (Western Nile Delta, North Western Desert, Egypt)," Quat. Int., vol. 542, pp. 109–120, Mar. 2020.

[34] P. J. Moorad, "A comparative study of medicine among the Ancient races of the East: Egypt, Babylonia and Assyria," Ann. Med. Hist., vol. 9, pp. 155–167, 1937.

[35] B. Davis and J. Toivari, "A Letter of Reproach (O. DeM 314). Corruption in the Administration of the Washing Service at Deir El-Medina," in Deir El-Medina in the Third Millennium AD. A Tribute to Jac. J. Janssen, R. J. Demaree and A. Egberts, Eds. 2000, pp. 65–77.

[36] S. Isserow and H. Zahnd, "Chemical knowledge in the old testament," J. Chem. Educ., vol. 20, no. 7, pp. 327–335, Jul. 1943.

[37] H. B. Routh, K. R. Bhowmik, L. C. Parish, and J. A. Witkowski, "Soaps: From the phoenicians to the 20th century —A historical review," Clin. Dermatol., vol. 14, no. 1, pp. 3–6, Jan. 1996.

[38] K. L. Konkol and S. C. Rasmussen, "An Ancient Cleanser: Soap Production and Use in Antiquity," 2015, pp. 245–266.

[39] N. Everett, The Alphabet of Galen: Pharmacy from Antiquity to the Middle Ages: Pharmacy from Antiquity to the Middle Ages: A Critical Edition of the Latin Text Paperback. Toronto, Canada: University of Toronto Press, 2012.

[40] A. S. . Davidsohn, "Soap and detergent," Encyclopedia Britannica, 2020. [Online]. Available: https://www.britannica.com/science/soap. [Accessed: 23-Mar-2021].

[41] M. Levey, "Mediaeval Arabic Bookmaking and Its Relation to Early Chemistry and Pharmacology," Trans. Am. Philos. Soc. New Ser., vol. 52, no. 4, pp. 1–79, 1962.

[42] M. Cook, "The Leblanc Soda Process. A Gothic Tale for Freshman Engineers," Chem. Eng. Educ., vol. 32, no. 2, pp. 133–137, 1998.

[43] D. Dungworth, "The Value of Historic Window Glass," Hist. Environ. Policy Pract., vol. 2, no. 1, pp. 21–48, Jun. 2011.

[44] H. O. Ogunsui and C. A. Akkinawo, "Quality Assessment of Soaps Produced from Palm Bunch Ash-Derived Alkali and Coconut Oil," J. Appl. Sci. Environ. Manag., vol. 16, no. 4, pp. 363–366, 2012.

[45] M. D. Torres and J. Seijo, "By-Products from the Chestnut Industry Used to Produce Natural Potassium Soaps: Physicochemical Properties," J. Surfactants Deterg., vol. 19, no. 2, pp. 381–387, Mar. 2016.

[46] O. Atolani et al., "Green synthesis and characterisation of natural antiseptic soaps from the oils of underutilised tropical seed," Sustain. Chem. Pharm., vol. 4, pp. 32–39, Dec. 2016.

[47] A. J. Shortland and T. Rehren, "Glass," in Archaeological Science, M. P. Richards and K. Briton, Eds. Cambridge University Press, 2020, pp. 347–364.

[48] T. Rehren and D. Rosenow, "Three Millennia of Egyptian Glassmaking," in Mobile Technologies in the Ancient Sahara and Beyond, Cambridge University Press, 2020, pp. 423–450.

[49] J. E. Shelby, Introduction to Glass Science and Technology, 3rd Editio. Cambridge: Royal Society of Chemistry, 2020.

[50] A. E. Marshall, "An Assyrian text on glass manufacture," J. Chem. Educ., vol. 10, no. 5, p. 267, May 1933.

[51] C. M. Jackson, S. Paynter, M.-D. Nenna, and P. Degryse, "Glassmaking using natron from el-Barnugi (Egypt); Pliny and the Roman glass industry," Archaeol. Anthropol. Sci., vol. 10, no. 5, pp. 1179–1191, Aug. 2018.

[52] C. M. Jackson, C. A. Booth, and J. W. Smedley, "Glass by Design? Raw Materials, Receipes and Compositional Data," Archaeometry, vol. 47, no. 4, pp. 781–795, Nov. 2005.

[53] J. Hemderson, "The Raw Materials of Early Glass Production," Oxford J. Archaeol., vol. 4, no. 3, pp. 267–291, Nov. 1985.

[54] G. Dardeniz, "Was Ancient Egypt the Only Supplier of Natron? New Research Reveals Major Anatolian

Deposits," in Anatolica XLI (Volume 41), 2015, pp. 191–202.

[55] Z. Cílová and J. Woitsch, "Potash – a key raw material of glass batch for Bohemian glasses from 14th–17th centuries?," J. Archaeol. Sci., vol. 39, no. 2, pp. 371–380, Feb. 2012.

[56] G. H. Barfod, I. C. Freestone, A. Lichtenberger, *R. Raja*, and H. Schwarzer, "Geochemistry of Byzantine and Early Islamic glass from Jerash, Jordan: Typology, recycling, and provenance," Geoarchaeology, vol. 33, no. 6, pp. 623–640, Nov. 2018.

[57] A. B. Babalola, L. Dussubieux, S. K. McIntosh, and T. Rehren, "Chemical analysis of glass beads from Igbo Olokun, Ile-Ife (SW Nigeria): New light on raw materials, production, and interregional interactions," J. Archaeol. Sci., vol. 90, pp. 92–105, Feb. 2018.

[58] M. S. Tite, A. Shortland, Y. Maniatis, D. Kavoussanaki, and S. A. Harris, "The composition of the soda-rich and mixed alkali plant ashes used in the production of glass," J. Archaeol. Sci., vol. 33, no. 9, pp. 1284–1292, Sep. 2006.

[59] W. B. Stern and Y. Gerber, "Potassium-Calcium Glass: New Data and Experiments*," Archaeometry, vol. 46, no. 1, pp. 137–156, Feb. 2004.

[60] J. Smedley, C. M. Jackson, and C. A. Booth, "Back to the Roots: the Raw Materials, Glass Receipes and Glassmaking Practices of Theophilus.," in The Prehistory and History of Glassmaking Technology, Westerville, Ohio: The American Ceramic Society, 1998, pp. 145–165.

[61] A. Shortland, L. Schachner, I. Freestone, and M. Tite, "Natron as a flux in the early vitreous materials industry: sources, beginnings and reasons for decline," J. Archaeol. Sci., vol. 33, no. 4, pp. 521–530, Apr. 2006.

[62] A. J. Shortland, P. Degryse, M. Walton, M. Geer, V. Lauwers, and L. Salou, "The Evaporitic Deposits of Lake Fazda (Wadi Natrun, Egypt) and Their Use in Roman Glass Production," Archaeometry, vol. 53, no. 5, pp. 916–929, Oct. 2011.

[63] T. Rehren and M. Brüggler, "The Late Antique glass furnaces in the Hambach Forest were working glass - not making it," J. Archaeol. Sci. Reports, vol. 29, p. 102072, Feb. 2020.

[64] T. Rehren, "A review of factors affecting the composition of early Egyptian glasses and faience: alkali and alkali earth oxides," J. Archaeol. Sci., vol. 35, no. 5, pp. 1345–1354, May 2008.

[65] K. Kanungo and R. H. Brill, "Kopia, India's First Glassmaking Site: Dating and Chemical Analysis," J. Glass Stud., vol. 51, pp. 11–25, 2009.

[66] P. Degryse, R. B. Scott, and D. Brems, "The archaeometry of ancient glassmaking: reconstructing ancient technology and the trade of raw materials," Perspective, no. 2, pp. 224–238, Dec. 2014.

[67] Pliny, Natural History, vol. 8. Cambridge, Mass. and London: Loeb Classical Library, 330, 1963.

[68] Y. Shindo, "Islamic Lustre-Stained Glass from Raya between the Ninth and Tenth Centuries," in AIHV du 16e Congrès. London 2003, 2005, pp. 174–177.

[69] E. Shovelton, "Ceramic dish," Discover Islamic Art, Museum With No Frontiers, 2021. [Online]. Available: http://islamicart.museumwnf.org/database_item.php?id=object;ISL;uk;Mus01;14;en&cp. [Accessed: 29-Jun-2021].

[70] N. Kato, I. Nakai, and Y. Shindo, "Transitions in Islamic plant-ash glass vessels: on-site chemical analyses conducted at the Raya/al-Tur area on

the Sinai Peninsula in Egypt," J. Archaeol. Sci., vol. 37, no. 7, pp. 1381–1395, Jul. 2010.

[71] R. W. Sowers, "Stained glass," Encyclopedia Britannica. 2021.

[72] H. Louw, "The Development of the Window," in Windows, Routledge, 2015, pp. 7–96.

[73] Wikipedia – the Free Encyclopedia, "Medieval stained glass" [Online]. Available: https://en.wikipedia.org/wiki/Medieval_stained_glass. [Accessed: 27-Jule-2021].

[74] Historic England, "Traditional Windows: Their Care, Repair and Upgrading," p. 74, 2014.

[75] "Verrerie . Verrerie en Bois, ou Grande Verrerie a Vitres ou en Plats. Eds. Denis Diderot and Jean le Rond D'Alembert. Paris: Chez Briasson, David & Le Breton-1767," University of Chicago: ARTFL Encyclopédie Project. University of Chicago. [Online]. Available: https://artflsrv03.uchicago.edu/philologic4/encyclopedie1117/navigate/27/31/. [Accessed: 29-June-2021]

[76] A. Silvestri and A. Marcante, "The glass of Nogara (Verona): a 'window' on production technology of mid-Medieval times in Northern Italy," J. Archaeol. Sci., vol. 38, no. 10, pp. 2509–2522, Oct. 2011.

[77] S. Cagno, L. Favaretto, M. Mendera, A. Izmer, F. Vanhaecke, and K. Janssens, "Evidence of early medieval soda ash glass in the archaeological site of San Genesio (Tuscany)," J. Archaeol. Sci., vol. 39, no. 5, pp. 1540–1552, May 2012.

[78] S. C. Rasmussen, "Advances in 13th Century Glass Manufacturing and Their Effect on Chemical Progress," Bull. Hist. Chem., vol. 33, no. 1, pp. 28–33, 2008.

[79] O. Schalm, K. Janssens, H. Wouters, and D. Caluwé, "Composition of 12–18th century window glass in Belgium: Non-figurative windows in secular buildings and stained-glass windows in religious buildings," Spectrochim. Acta Part B At. Spectrosc., vol. 62, no. 6–7, pp. 663–668, Jul. 2007.

[80] K. Janssens, S. Cagno, I. De Raedt, and P. Degryse, "Transfer of Glass Manufacturing Technology in the Sixteenth and Seventeenth Centuries from Southern to Northern Europe: Using Trace Element Patterns to Reveal the Spread from Venice via Antwerp to London," in Modern Methods for Analysing Archaeological and Historical Glass, K. Janssens, Ed. Oxford, UK: John Wiley & Sons Ltd, 2013, pp. 537–562.

[81] Encyclopædia Britannica, "The making of broad glass, from an engraving of a German glassworks, 1865," Encyclopædia Britannica. [Online]. Available: https://www.britannica.com/topic/glass-properties-composition-and-industrial-production-234890/History-of-glassmaking#/media/1/234890/261. [Accessed: 29-June-2021].

[82] *W. Cooper*, The crown glass cutter and glazier's manual. Edinburgh: Oliver & Boyd, 1835.

[83] The Editors of Encylopedia, "Kelp," Encyclopedia Britannica, 2020. [Online]. Available: https://www.britannica.com/science/kelp. [Accessed: 06-Apr-2021].

[84] J. Wisniak, "Sodium carbonate — From natural resources to Leblanc and back," Indian J. Chem. Technol., vol. 10, no. 1, pp. 99–112, 2003.